Aspects of Geography

General Editors: T. H. Elkins and Keith

Hydrology:

LONDON BOROUGH OF ENFIELD
LIBRARY SERVICES

This book to be RETURNED on or before the latest date stamped unless a renewal has been obtained by personal call, post or telephone, quoting the above number and the date due for return.

First published 1979
Reprinted 1979

Published by
Macmillan Education Limited
Houndmills Basingstoke Hampshire RG21 2XS
and London
Associated companies in Delhi Dublin
Hong Kong Johannesburg Lagos Melbourne
New York Singapore and Tokyo

Printed in Hong Kong

British Library Cataloguing in Publication Data

Newson, Malcolm David
    Hydrology. Measurement and application. — (Aspects
of Geography).
    1. Hydrology –Measurement
    I. Title   II. Series
    551.4'8   GB680

    ISBN 0-333-24366-8

Contents

Acknowledgements

The author and publishers wish to thank the following who have kindly given permission for use of copyright material:

J. Allen Cash, for permission to reproduce the cover photograph.

Edward Arnold (Publishers) Limited for drawings from *Drainage Basin Form and Process* (1973) by K. J. Gregory & D. E. Walling

Professor R. J. Chorley for drawings from 'Water, Earth & Man' (1969)

The Controller of Her Majesty's Stationery Office for two maps from the Meteorological Office and *Irrigation in Britain* (1962)

The Geographical Association for figures from *Hydrology for Schools* (1975) by D. Weyman & C. Wilson and map from 'Rethinking Our Approach to Water Supply Provision' by J. Rees featured in *Geography*, Volume 61, November 1976

McGraw-Hill Book Company (UK) Limited for drawings from 'Principles of Hydrology' 2nd Edition by R. C. Ward

Oxford University Press for figures from 'Flooding and Flood Hazard' (1975) by M. D. Newson.

The author would also like to thank Mrs S. Hill for help in producing this text.

Preface

On average water vapour stays in the atmosphere for just eleven days, so that the atmospheric engine, driven by the sun, is continually dumping very large volumes of precipitation on the land masses. Almost as quickly, much of this water finds its way back to the ocean down our streams and rivers. Some, of course, is evaporated back to the atmosphere, and some is stored in lakes, ice-caps and glaciers, or as soil moisture or groundwater. This water balance is studied by hydrologists, and their approach and the measurements they make are outlined in this book.

Anyone living in a humid climate will be near streams and rivers of all sizes, with their constantly changing behaviour as discharge varies with the pattern of rainfall. This means that hydrology can be pursued as a very practical field study, and this book includes a number of suggestions about what to measure and how it can be done with relatively simple equipment.

T. H. ELKINS
KEITH CLAYTON

Preface

Basic concepts in hydrology

The hydrological cycle and scales of studying it

Part of the circulation of energy in the atmosphere which produces the different climates of the world involves water in its three main states: gas, liquid, solid. The basic laws of thermodynamics tell us that matter and energy are interchangeable and form part of a constantly balancing equation with no net losses or gains, only circulation. The circulation of water between atmosphere, land and ocean is known as the hydrological cycle. It can be studied at a variety of scales, *global*, as part of the assessment of Man's finite resources, *national*, for use by government agencies, *catchment*, for research on a single river basin and finally at what we might call an *experimental* scale, where physical processes are studied in great detail.

We begin here with the global scale since it provides a good illustration of the principle of a *cycle* of activity and the quantities of water involved. However, the catchment scale is the dominant one in both research and education and, after an introduction in this chapter, it forms the focus of much of the rest of the book.

The global scale

At any instant of time the total world store of water may be located in a variety of situations, and in a variety of forms. The major locations and the proportions of total water stored in them are shown in table 1.

Notice how small is the part of the total available to us in rivers. Of the world's fresh water (only 2·5 per cent of the total water) most is locked up as polar ice caps – enough for 1000 years of average world river flow.

Total volume	1384 million km³
World ocean	97·6%
Ice caps and glaciers	1·9%
Groundwater	0·5%
Soil moisture	0·01%
Lakes and rivers	0·009%
The atmosphere	0·0001%

Table 1 Major locations of global water resources, and the proportions of the total water volume stored in them

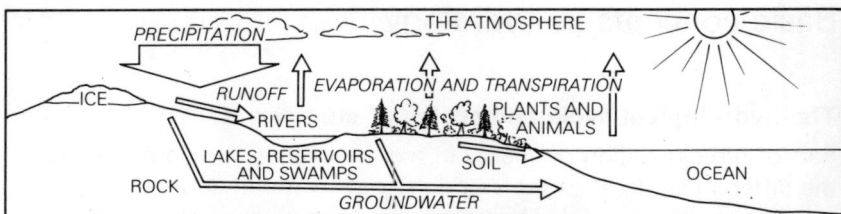

THE ATMOSPHERE 13 000 km^3, 8 – 10 days   RIVERS 1700 km^3, 2 weeks
LAKES RESERVOIRS AND SWAMPS 125 000 km^3, years   PLANTS AND ANIMALS 700 km^3, 1 week
SOIL 65 000 km^3, 2 weeks – 1 year   ROCK 7 million km^3, days – thousands of years
ICE 26 million km^3, thousands of years   OCEAN 1370 million km^3, thousands of years

Figure 1   **The hydrological cycle as a global concept, showing STORAGES and *FLOWS*. The size of the storages and the average time spent in them by each molecule of water are tabulated.**

The locations of water shown in table 1 represent the nodal positions in the system depicted in figure 1. We next examine the nature of the circulatory movements between them. We know that the evaporation of water from the oceans, smaller open water bodies and vegetated land areas permits the Earth's wind systems to redistribute the resulting water vapour. Condensation in the form of rain or snow takes it to earth where it enters the soil or flows direct to rivers, or lakes. Much, of course, falls directly into the oceans. A substantial volume of the precipitated water may become 'lost' by being evaporated from the land or ocean directly, or by being taken up by biological systems of plants and animals. In reality none is lost – it is merely recycled through the atmosphere to fall again as precipitation.

The turnover time of moisture in each of the temporary stores of the hydrological cycle varies. Judging from the average time spent by a particle of water in each major nodal position in the cycle (figure 1) the most dynamic part of the cycle is the atmosphere, followed by rivers and lakes. The hydrology of the *atmosphere* is the province of the meteorologist; meteorology is an important background to the earth's water resources. Since the oceans are so vast and largely uninhabited by Man, their hydrology is only tackled by those working on attempts to measure the hydrological balance of the world.

The small amount of water which spends such a short time over the continental areas of the globe is the subject of this text and may be referred to as the *land phase* of the hydrological cycle. However, the land phase is the most diverse in spite of the relatively small quantities involved. Since Man is deeply involved with this phase for practical reasons, national boundaries obviously play some part in scaling hydrological research, the individual national activities being coordinated where possible by United Nations agencies as, for instance, during the International Hydrological Decade, 1965–75. However, within each nation the usual scale of study is the river basin (*drainage basin* or *catchment*). As the unit from which the surplus rainfall of the area flows, it is the most convenient unit for our routine measurements.

The catchment scale

In the British Isles the term 'catchment' is more commonly used than 'drainage basin' to describe the fundamental unit of hydrological study. 'Drainage basin' is more commonly used in the United States and tends also to be preferred by geomorphologists. The natural catchment is a complicated unit, having a cover of soils and vegetation and a foundation of various rock types. It is bounded by the watershed line which separates slopes which shed runoff to one catchment from those contributing to its neighbour. The geomorphologist and hydrologist work together in taking the catchment as the unit in which inputs of rainfall and solar energy result in outputs of evaporation, river flow and fluvial erosion.

The balancing of quantities implicit in the global hydrological cycle is also a feature of the catchment cycle. The hydrologist attempts to measure the various parts of the catchment *water balance* (figure 2).

Phrased as an equation the water balance may be written as:

river flow = precipitation − evapotranspiration ± changes in soil moisture and groundwater

or, put in a more general way:

output = input − losses ± storage changes

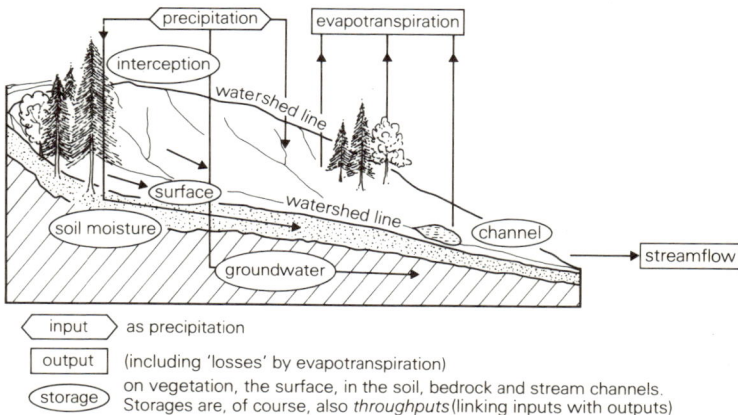

Figure 2 The catchment water balance, adding detail to the global picture for the case of a humid temperate rural area

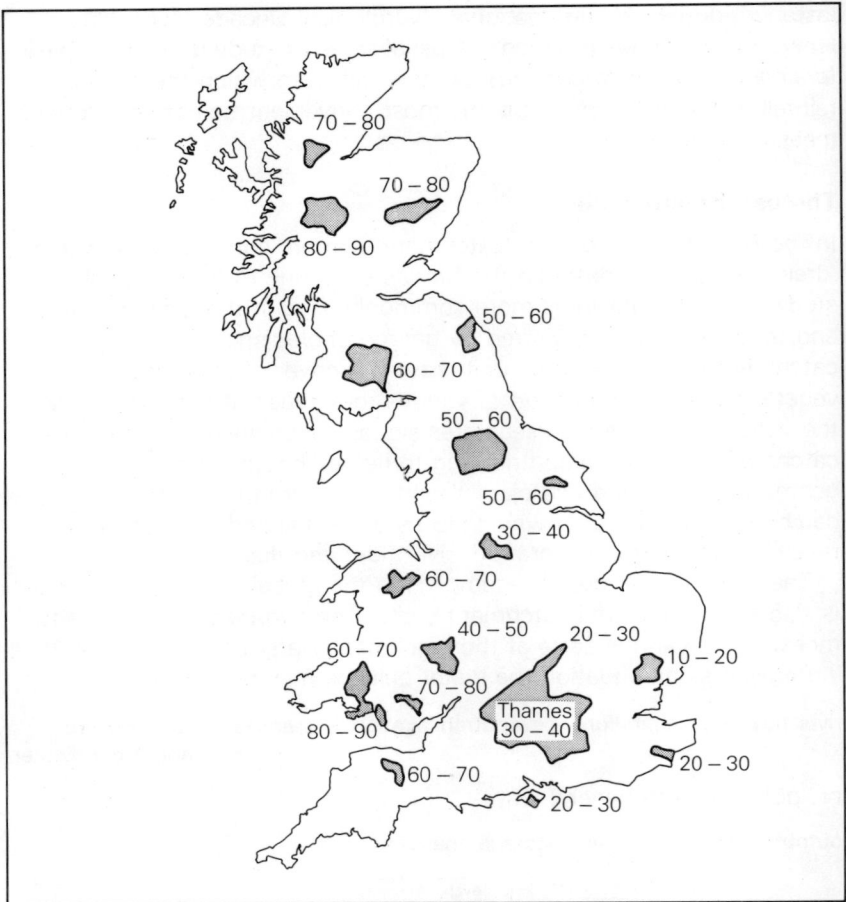

Figure 3 The runoff coefficient for selected British catchments (figures show annual river flow as percentages of annual rainfall)

By taking periods over which the changes in storage are negligible, for example, from the end of one summer to the end of the next summer, or over a number of years, we can concentrate on river flow, precipitation and evapotranspiration (although it will be seen on pp. 26–32 that storages can be measured).

By putting figures to these terms for two catchments, one in East England and one in West Wales we observe the range of values.

East England   307 mm = 787 mm − 480 mm ± 0
West Wales   2320 mm = 2794 − 474 mm ± 0

(Note that while rainfall increases east to west by a factor of 3·5, losses change less, even decreasing; thus runoff is much greater in the west, by a factor of 7·5.)

By expressing river flow as a percentage of precipitation we can generalise over the whole of Britain (see figure 3). This percentage is known as the *runoff coefficient* and its range shows again the dominant position of the north and west in generating 'surplus' water from precipitation. This factor is an important one in British water supply, especially because concentrations of population tend to occur where rainfall and runoff are least, and hence must be supplied from areas with larger surpluses (see figure 36).

Obviously, from the contrast in the two water balance equations and the map of runoff coefficients, climatic influences (rainfall and evaporation) are important for runoff. Physiographic influences are also strong. The description of the size, shape, slope and surface properties (soil, land-use, vegetation) of river catchments is an important part of hydrological analysis, since these properties determine the way in which meteorological inputs are translated into streamflow outputs. What are the processes which produce *runoff* from rainfall? Our knowledge of these, summarised below, has largely come from catchment studies and from work at the experimental scale within catchments.

Rainfall and runoff processes

The routes by which excess rainfall reaches the stream, how much does so, and at what time during a rain storm, have only recently been elaborated. Since the catchment consists of slopes and channels there are two main phases of runoff. In the *slope phase* rainfall first encounters the vegetation cover, if one exists. Since the vegetation 'canopy' of leaves intercepts some rainfall, not all of it reaches the soil surface. *Interception* is a recently-studied process and the proportion of rainfall intercepted varies between 19 per cent and 50 per cent for conifers in moderate rainfall. Over 80 per cent of very light rainfall may be intercepted by the canopy of a spruce forest. Lower proportions are recorded for grass or crops or in heavier rainfall.

The *net rainfall*, which reaches the soil surface, behaves in two possible ways. Where the soil is already soaked by rain or highly compacted (for example clays compressed by agricultural machinery) or if the rainfall is very heavy, *surface runoff* will occur; rain flows downslope across the soil surface and either collects in depressions or reaches the nearest small channel. Obviously where the soil is not bound by vegetation this process results in erosion and even in gullying. In most soils, water *infiltrates* the soil profile and travels vertically through it until a less permeable horizon is reached; at this boundary, part will then move downslope in the soil as *throughflow*. The infiltration/throughflow circulation can be very rapid, especially if the upper soil is well aerated, or cracked, or if animal burrows or root channels cross it. Deeper percolation beneath the soil and into the local drift deposit or bedrock produces a generally slower route to the stream channel via the *groundwater* system. Groundwater contributions do not generally form a high percentage of runoff from an individual storm, rather contributing to the background *base-flow*. The timing of runoff contributions during a storm are shown in figure 4.

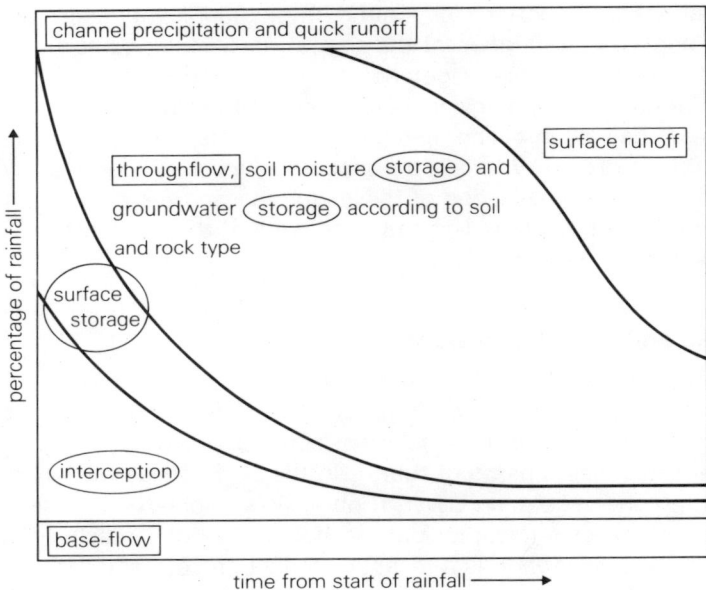

Figure 4 Timing of storages and runoff routes during a storm. Box symbols are as for figure 2, representing storages (throughputs) and outputs; e.g. at the start of the storm, interception and surface drainage dominate. By the end, these are small in comparison with surface runoff, throughflow etc., and the stream is in flood.

Much soil water enters the roots of plants and is evaporated at their leaves as part of *transpiration*. *Evaporation* can also act on water held at the soil surface in a wet soil (and draw up further supplies to that surface by capillary forces), on water held by interception on the canopy or, of course, on water detained in puddles, pools and lakes or in rivers themselves.

In the stream *channel phase* of runoff the hydrologist is examining the energy relationships of a body of water moving under gravity downstream but which dissipates a variable amount of gravitational energy in overcoming the friction of the bed and banks. The branch of physics dealing with this topic is hydraulics.

Runoff processes in space and time

The biggest conceptual problem in dealing with runoff from a catchment is that the activity of the above processes varies spatially over the catchment area, introducing *delays, storages, losses* and *flows* which then produce the pattern in time of the stream's response.

When rainfall begins, only that very small fraction which falls directly into the streams of the catchment will definitely form runoff. It will be joined, after a short delay, by surface runoff from those areas of soil which were already saturated before rain started (for example, the base of slopes and wet hollows on slopes). This saturated area expands if the storm goes on; surface runoff may be produced in sufficient amounts to initiate small channels. Throughflow through the soil takes a variable time to reach the channel and its contribution may be mainly to expand the saturated *'contributing area'*.

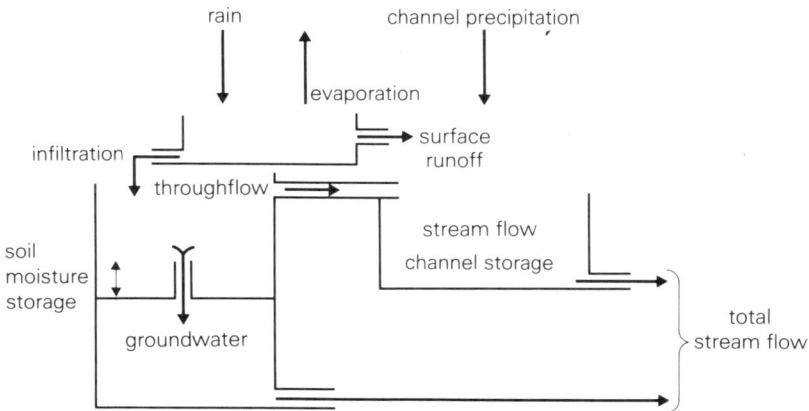

Figure 5 An elementary runoff model. Such a model can be constructed in the classroom from plastic bottles and tubing, and made to work by pouring water gently into the top container.

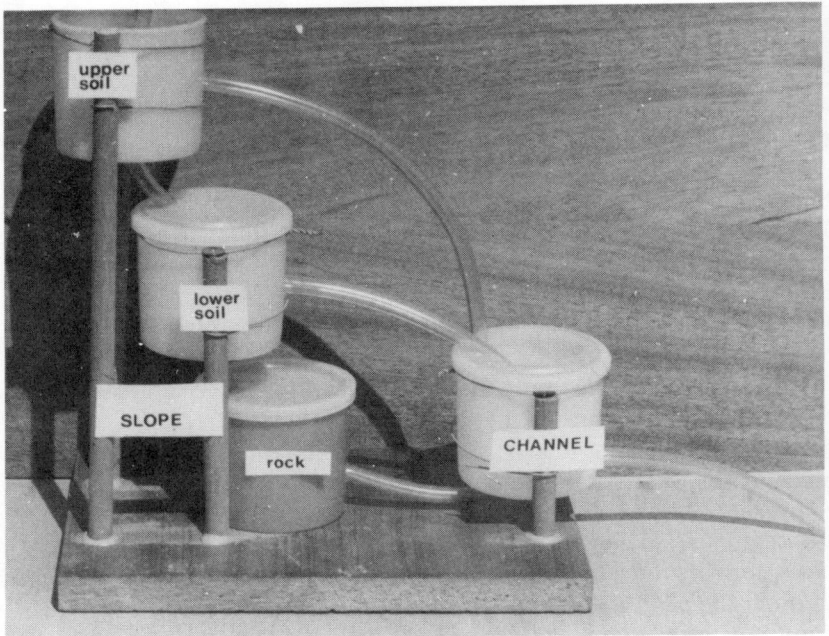

Figure 6 A simple hardware model of the catchment as a series of storages and flows

In the channel the contributions from slopes will form the *hydrograph*, that distinctive rise and fall of river level which marks the passage of a flood. (This concept is considered further in a later chapter, as it is important in flood analysis.)

After the storm finishes, the hydrograph moves downstream as a *flood wave*, receiving further contributions if the storm moves with it but becoming flatter and longer because of the hydraulic effects of the channel and storage in pools or flooded fields. Between storms, redistribution of the water occurs: intercepted water is evaporated from the vegetation canopy, percolation and slower throughflows continue to operate and plants take up water through their roots (transpiration largely ceases during wet periods). River discharge decreases as a *recession curve* and the gradual reduction of this curve is part of drought analysis.

As soon as the hydrologist is confident that he understands the processes listed above he can begin to make a *conceptual model* of the catchment hydrology, incorporating his views of the way in which the catchment produces runoff. Since the rainfall is subjected to storages, some of which involve losses (for example, interception, soil moisture movements, the stream channel itself) and delays, all these can be quantified by measurements and a mathematical model constructed. It is also easy to build a simple hardware model of the catchment in plastic in which the storages are containers and the linking flows are tubes (figures 5 and 6).

Measuring rainfall and streamflow

Whether to quantify our mathematical models of catchments or to collect data for statistical analysis, measurement is still the front line of hydrological work.

Rainfall

The large range of rainfall amounts across the British Isles has already been referred to. Within these regional variations there are local variations with altitude (rainfall increasing with height above sea-level), exposure to rain-bearing winds, the area covered by storms, urban influences and so on. Therefore, for almost any catchment study except the very small, a number of raingauges are required. This can involve two problems for the hydrologist: the performance of the individual raingauge in making *point measurements* and that of combining the various gauges in the raingauge network to give an *areal average*.

Point measurement

The raingauge is a deceptively simple instrument consisting of a standard-size *funnel* [127 mm (5 inches) diameter], *collecting can* or bottle and *outer protective container*, buried in the earth so that the funnel rim is level and raised 0·3m (1 foot) above ground surface. If precautions are taken to imitate the standard design, a home-made raingauge can be made from a plastic funnel and an empty wine bottle! (See figures 7a and 7b.)

The depth of water collected in the can or wine bottle is *not* the depth of rainfall. A separate, narrow measuring cyclinder is normally used, with graduations calculated by using the area of the funnel to convert 1 mm, 2 mm, 3 mm, etc. of rainfall to a volume of catch. An appreciation of depths, areas and volumes is essential in hydrology and calibrating your own raingauge and measuring cylinder is good practice.

There are several dangers involved in the standard design of raingauges. For instance, will the presence of the gauge itself influence the rainfall caught by causing *air turbulence* as the wind blows round it? Will rainfall falling at an angle be caught accurately by a level funnel? Will raindrops hit the funnel and *splash* off? Will the rain caught in the can or bottle be lost by *evaporation*? With careful selection of site most of these dangers can be averted. The one most difficult problem left is that of rainfall in mountainous or hilly areas which is invariably accompanied by strong winds. It is now proven

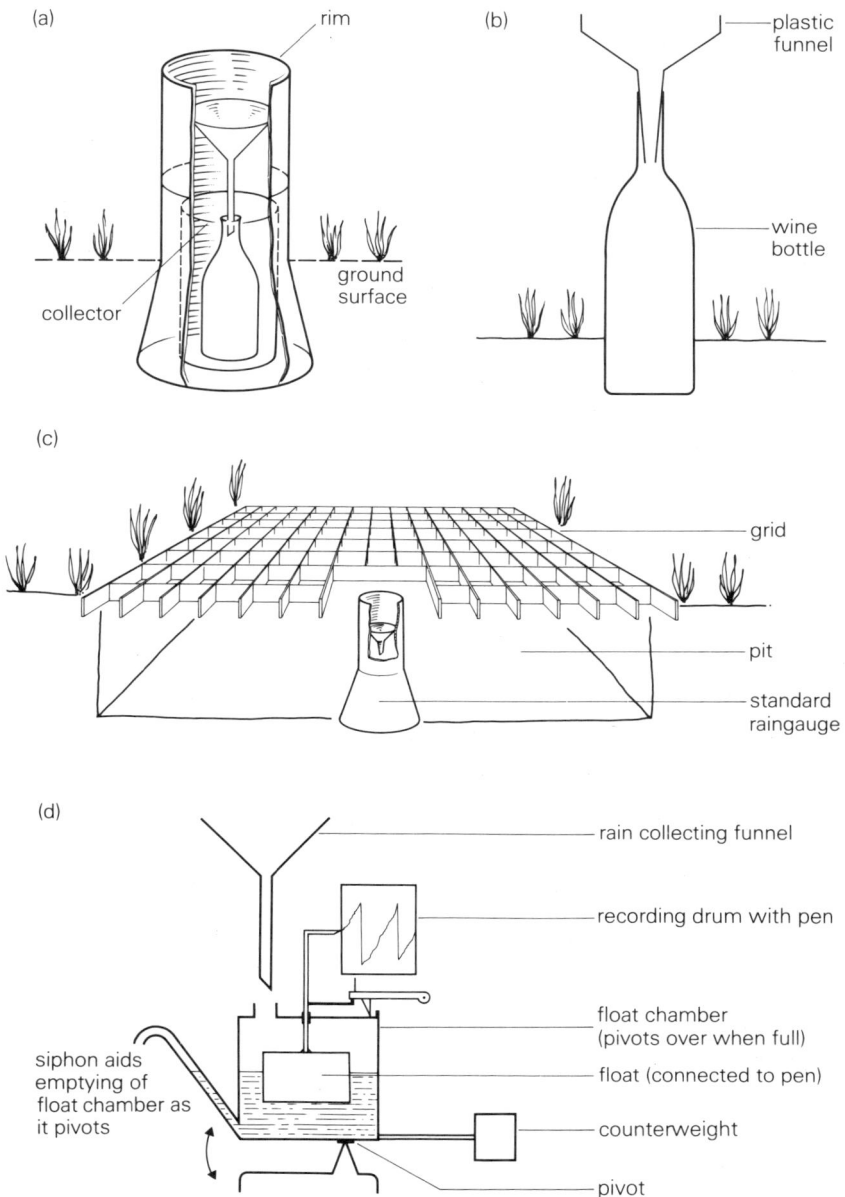

Figure 7 Raingauges (a) Standard British Meteorological Office daily raingauge (b) Home-made funnel gauge (c) Ground-level raingauge (using standard gauge) (d) Principles and inner workings of the 'Dines' tilting siphon rainfall recorder (these inner workings are housed inside a large version of the standard gauge)

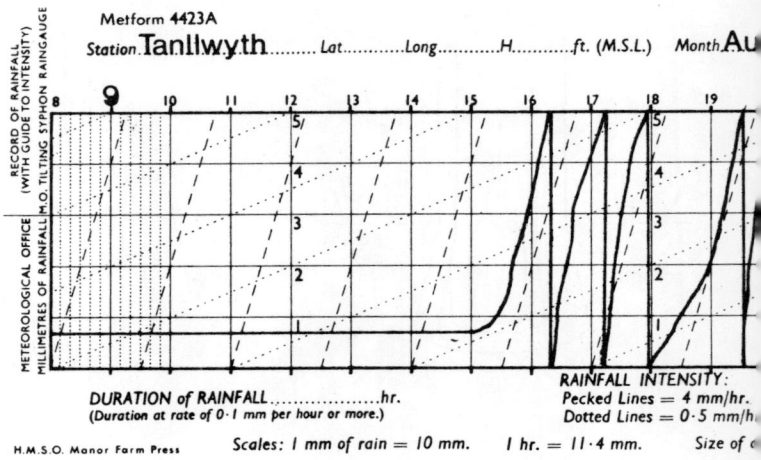

Figure 8 Daily-read Dines recorder chart for Tanllwyth (Moel Cynnedd), Severn catchment

that, in such conditions, air turbulence around the standard gauge does result in under-catching by the funnel. This has been established with wind-tunnel tests and by developing a raingauge which operates at *ground level*, surrounded by a screen to stop splashing in of extra rain from surrounding vegetation. Such ground-level gauges (figure 7c) are expensive to install and consequently most mountain raingauges are either carefully sited to avoid gusty winds or protected by a low wall of turf at a radius of 1·5 m from the gauge itself.

Most raingauges are read once-daily at nine o'clock in the morning. Care should be taken in emptying the contents into the graduated cylinder – record catches have occasionally been lost by spillage during this operation! The record British daily catch is 275 mm, which, for a standard raingauge funnel, means nearly 3·5 litres of water to remove carefully from the gauge to the calibrated rain-measure. The measured catch is accorded to the day before; thus, the rainfall for the 12th November is actually that emptied from the gauge at 0900 GMT on 13th November.

Daily attendance would not be possible at all the United Kingdom's 6864 raingauges – many are in remote spots. Consequently some are built with large enough containers ('*storage* gauges') to hold a month's catch and are read on the first day of the next month.

Even with daily or monthly rainfall totals we are not in a position to analyse the response of rivers to rainfall in any but a most general way. Most rivers rise and fall much more rapidly and there is a need

12

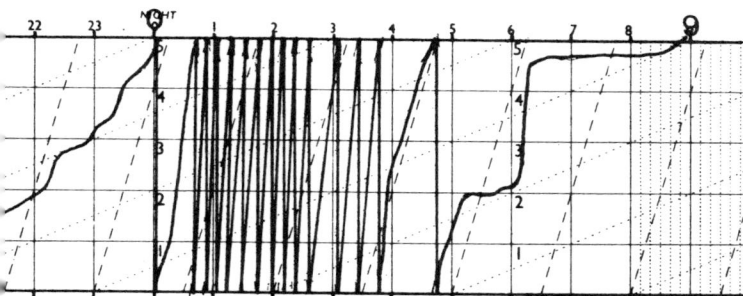

TOTAL by CHECK GAUGE 9h to 9h..**100.6**..........mm.
TOTAL by RAIN RECORDER...........**104.4**..........mm.

89 mm.   Margins: Left, 9 mm.   Right, Nil.   Upper, 21 mm.   Lower, 18 mm.

to obtain some information on the rate of rainfall (*rainfall intensity*) during at least hourly periods. Rather than sending out the observer once an hour, or more frequently, a variety of timing mechanisms have been incorporated in raingauges; these are then known as *rainfall recorders*.

The basic type of rainfall recorder consists of a large version of the standard gauge but it contains a collecting chamber fitted with a float (figure 7d). As rain falls and the chamber fills, the float rises and a lightweight pen attached to the top marks a line, rising up a *paper chart*. The chart is fixed to a cylindrical drum, driven round by clockwork. As the drum goes round and the rain keeps falling, the pen slants up the chart (figure 8). At the top the chamber is filled with enough rain to tilt over on its pivot and all its contents siphon out of the gauge. It then returns to level and the pen now rests at the base of the chart again. One complete cycle measures 5 mm of rainfall (the float chamber's dimensions incorporate the raingauge calibration) and in heavy storms the pen quickly performs several rises and falls. During dry periods a horizontal line is drawn round the chart. Charts are most commonly designed to be changed each day although clocks and charts to exist for weekly or longer periods of operation.

Obviously the arrangement described above is a delicate one and is upset easily. For instance freezing of the float chamber is common in winter; this is generally prevented by adding known amounts of antifreeze solution to the chamber (and subtracting the amount from the rainfall caught) or by heating the gauge with a light bulb if a supply of electricity is convenient.

Since rainfall recorders occasionally break down and because some rainfall is not recorded during the time taken to empty the chamber, each recorder site is normally equipped with a standard raingauge as a *check gauge* (see figure 9). Not every standard raingauge is, however, equipped with a recorder, there being only 981 rainfall recorders in the United Kingdom official network.

The computer era has resulted in moves to collect hydrological data in a form which is immediately convenient for computer processing. This *digital data* is achieved for rainfall by a change in the recorder mechanism. Rain from the funnel is directed into *tipping buckets*. Each gauge is equipped with two small collecting buckets balanced either side of a knife-edge (figure 19). A millimetre of rain tips the bucket and brings the empty one of the pair into position under the funnel. The action of tipping produces an electronic impulse which can be recorded on a cassette tape.

Areal averages
Have we taken a representative sample of the general rainfall in our single gauge? Is the timing of the rainfall from our rainfall recorder accurate enough to apply to a wide area? The answers, of course, depend on the rainfall's own variability and the size of the area.

For hydrological analyses of catchments we tend to arrange a network of raingauges to produce a 'good coverage' (without, sometimes, being very precise about what this means – experience plays a big part in network design!). Frequently a large number of gauges are installed and, after a while, their data are analysed to see how many can be removed. In the west of Britain most rainfall is steady, general rain, resulting from frontal systems with a substantial orographic component; consequently a network organised to sample altitudinal zones in proportion to their contribution to the catchment would be suitable. In the east where convectional, thunderstorm, rainfall is commonly a significant part of annual totals, a close network is also necessary to record the localised outbreaks of this type of rainfall.

How can we estimate the areal rainfall as an average of the individual gauge catches, assuming that our network gives a 'good coverage' of the area in which we are interested? Two methods are shown in figure 10.

The simplest method is to calculate the average of the catches. However, usually some form of weighted average is taken. If enough raingauges are available one is able to *interpolate* values between their totals and draw in *isohyets* (lines of equal rainfall). The average catchment rainfall can then be calculated by weighting the isohyetal values by half the area to the next isohyet on either side.

Figure 9 Rainfall recorder (open for inspection) and, in the background, checkgauge of the standard type being read by observer

An even simpler method and a very popular one is that of Thiessen polygons. Here the individual raingauge total is taken to be representative of an area around it drawn in by marking off mid-points between all pairs of raingauges. Each of the network of gauges thus becomes surrounded by a polygon whose area is used to weight its total.

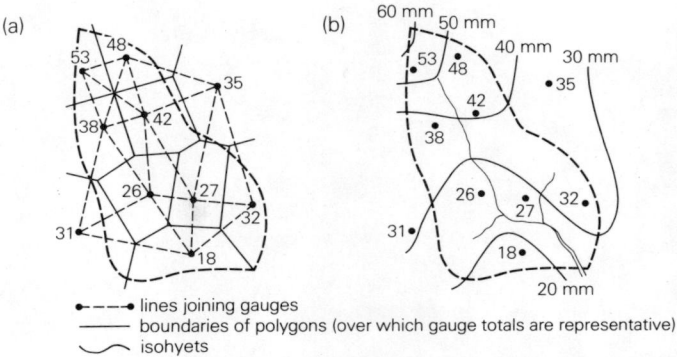

Figure 10 Averaging rainfall over a catchment area from point falls in ten gauges (dots) (a) Theissen Polygons (solid lines) (b) Isohyets. Note that rainfall totals gauged outside, but close to, the catchment can also be used. Copyright © 1975 McGraw-Hill Book Co. (UK) Ltd. From 'Principles of Hydrology' 2nd Edition by R. C. Ward. Reproduced by permission.

Other forms of precipitation

In some catchments and during certain seasons, forms of precipitation other than rainfall form an important input to the water balance. The most common example is that of solid precipitation in the form of snow. In the British Isles snow is a rather unpredictable aspect of hydrological operations during winter, occasionally falling and melting in large enough quantities to give widespread flooding, as for example in the spring of 1947.

The conventional raingauge is a poor recorder of snow, which is far more susceptible to wind turbulence than rainfall, even after it has fallen. Drifting around exposed raingauges is a common phenomenon and the ground-level configuration is little better.

On the continent, where snowfall forms a more significant input, raingauges are set on stakes with their rims at one metre above the ground to avoid drifts. In practice, snow measurements are best done by measuring depths over a large sample area with a rule and melting sample columns of known depth, removed for instance in a simple drainpipe corer or inverted raingauge funnel, to obtain the *water equivalent*. The water equivalent is the fraction of the snow mass occupied by water and may vary between around 10 per cent for fresh light snow to 30 per cent for 'old' snow or wet snow. The hydrologist, knowing the depth and water equivalent (or density) of the snow can calculate the equivalent amount of rainfall over the catchment.

Refinements to snow measurement do exist, such as snow stakes of various lengths from whose burial or exposure snow depth can be determined, ground-level photography in stereo pairs, or aerial photography.

Some effort is also made in hydrological research to measure, for example, dew, or the condensation of fog and mist on vegetation. Details of instruments for such work are given in the Meteorological Office *Observer's Handbook*.

To measure the net rainfall one could merely set up a raingauge beneath the canopy. However, the pattern of *throughfall* of rain is irregular and consequently long troughs are normally used in groups, each draining into a calibrated bin. Some water also reaches the ground as *stemflow*, down the trunks of trees for instance. This can be collected by fixing rubber traps round the trunk with polythene tubes leading the water off to a raingauge or bin (see figure 11). If the amount of interception is to be derived, a figure for the canopy level rainfall must be obtained by raingauging in a nearby clearing or hoisting a raingauge funnel on a pole above the canopy.

Figure 11 (left) Net rainfall measurement. Throughfall (left) and stemflow (right) measurements below a forest canopy.

Figure 12 (right) Small current meter in use. The operator stands well downstream of the propeller (shown above the surface here for clarity) and operates the counter box mechanism.

River flow

The problems of variations in space are not so great with river flow as with rainfall. One problem is immediately solved – where to install the gauge! So long as the catchment above the chosen point on the river can be defined on a map and does not leak (leakage can occur in very permeable rocks) a gauging site can be chosen. There are hydraulic constraints: a straight reach of the river is better than a bend and the flow should be as even as possible within the channel banks.

Why, then, do we not rely on gauging each major British river just before it enters the sea to get a gross impression of river flow on a national scale? The reason why tributary gauging is generally practised is variability in the flow regime's controlling factors of slope, soil, geology, vegetation and so on. As the relationships of river flows to catchment variables become known, it is possible that some gauging stations will become redundant. At present there are 1206 river gauging stations in the United Kingdom, most of them established in the last fifteen years.

Velocity and area

River flow is measured in units of cubic metres per second (m^3/s, or *cumecs*). These units result from the multiplication of the cross-sectional area of the river (m^2) and the velocity of its flow (m/s). Consequently, the fundamental method used to measure river flow (or *discharge*) is to survey the stream's width and depth at the chosen site and measure its velocity with a *current-meter* (figure 12).

Water surface width is comparatively easy to measure by survey but depth is made variable by irregular channel beds (especially in rocky streams). Therefore several measurements of depth are required across the stream to get an accurate figure for cross-sectional area (figure 13a). The same is true for measurements of velocity with the current-meter. The pattern of velocities is fairly regular if a straight reach is chosen and so a number of points are taken between the centre of the channel and the banks to get a sample of the decreasing velocity towards the banks (figure 13b(i)). To save sampling throughout the depth of the stream with the current-meter it has been calculated that average velocity occurs at 0·6 of the water's depth, measured from the surface down (figure 13b(ii)). Surface velocities are higher, so that if a float is used (oranges work well) instead of a proper meter, the surface velocities obtained should be multiplied by 0·8 to reduce them close to the average velocity.

In very small streams, or mountain rivers with very turbulent flows across irregular beds, the current-meter is unsuitable for measurement: velocity is too variable. Consequently a technique of

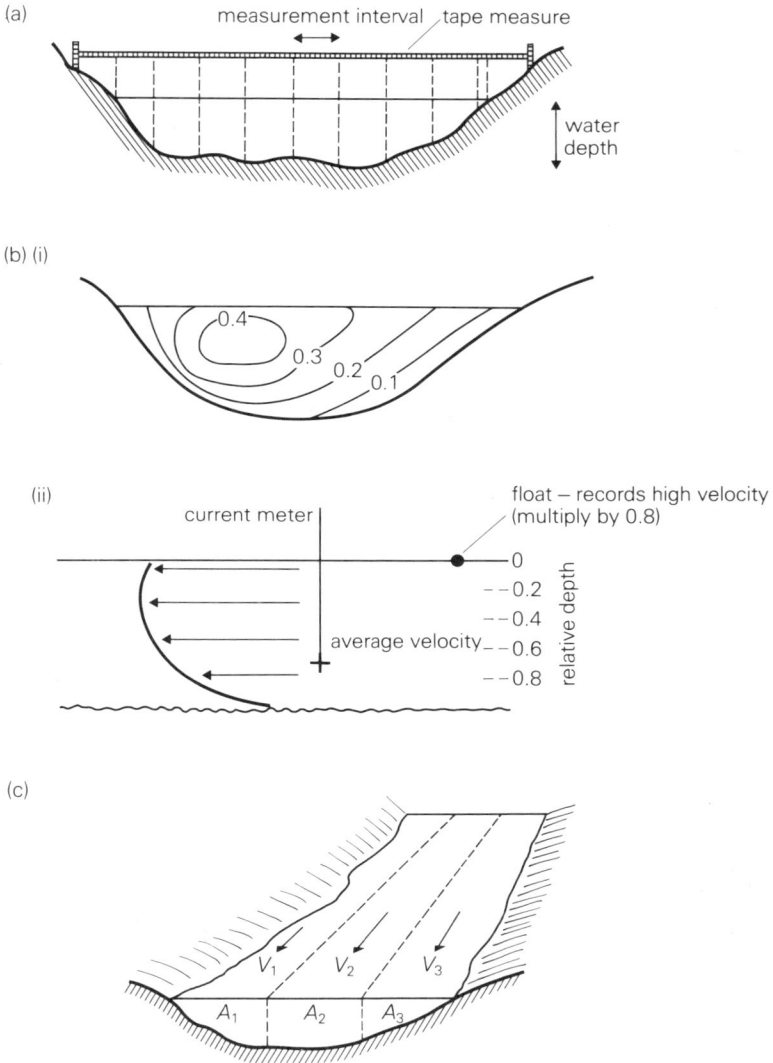

Figure 13 Velocity/area river gauging (a) Measurement of the cross-
sectional area of a river, taking vertical water depths at set intervals across
the channel (b) Variations in flow velocity in a river (i) Cross-sectional
variations shown by isovels (lines of equal velocity in m/s). This is not a good
cross-section for gauging. (ii) Vertical variations – the arrow length gives
relative velocity. Wire and cane floats are shown at various relative depths.
(c) Measurement of stream discharge: total discharge is the sum of velocity
times area ($V \times A$) for each lateral strip. You would use a current meter in the
middle of A_1, A_2 and A_3. Floats would be more difficult to use in three
discrete bands.

(a) the fieldwork

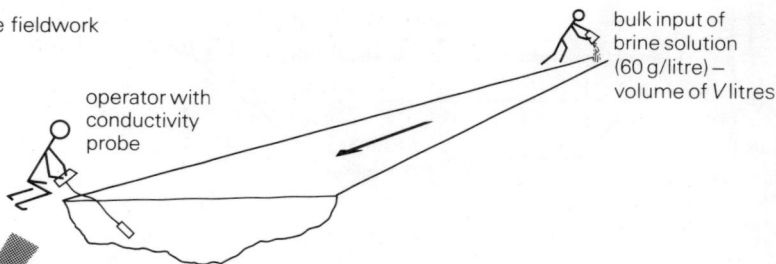

bulk input of brine solution (60 g/litre) — volume of V litres

operator with conductivity probe

(b) the plot of results

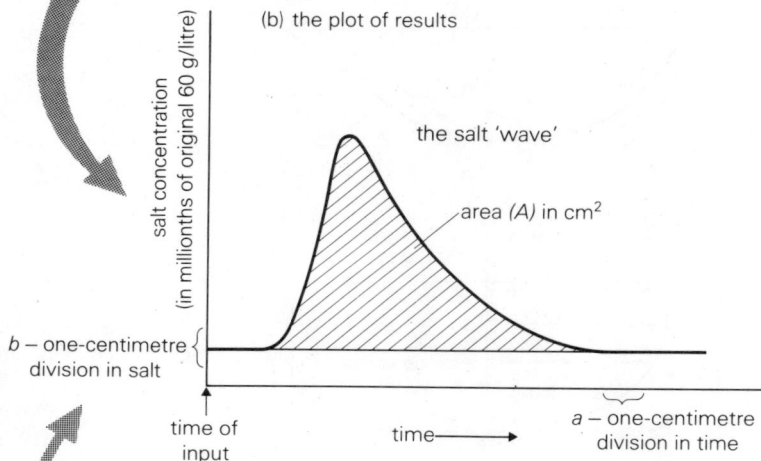

salt concentration (in millionths of original 60 g/litre)

the salt 'wave'

area (A) in cm²

b – one-centimetre division in salt

time of input

time ⟶

a – one-centimetre division in time

(c) the laboratory work to calibrate the probe

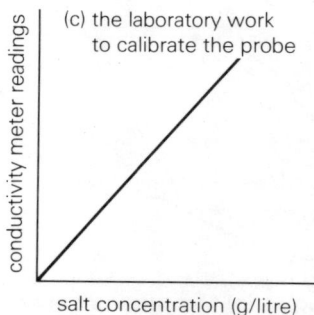

conductivity meter readings

$$\text{discharge} = \frac{V}{A \times a \times b}$$

(in litres per second)

(d) the equation

salt concentration (g/litre)

Figure 14   Dilution gauging with salt solution

gauging with a chemical is used. *Dilution gauging* uses a dose of concentrated chemical (for example, common salt solution) either in bulk or as a continuous stream. The dilution of this chemical in the stream is in fact aided by the mixing caused by turbulence and at a point downstream, where it is reckoned that mixing is complete, the wave of diluted chemical is sampled. If common salt is used, a simple electrical conductivity meter will monitor the salt concentrations. The time taken for a wave to pass (in the case of a bulk dose) will give the velocity of the stream and the dilution will give the volume of water. Thus discharge can be calculated. The method and calculations are illustrated in figure 14. Remember that chemicals added to streams may affect biotic systems and water supply.

Both current-metering and dilution gauging can be used for individual gaugings on streams but the aim is to set up stations which continuously record the river flow (just as the rainfall recorder backs up the raingauge). For this, several point gaugings are done at varying discharges (thirty or more) and the discharges plotted against river level read from a *gauge-board* secured to the channel bank. Using logarithmic graph-paper, the levels and discharges tend to plot on a straight line whose equation can be calculated. This *rating equation* (figure 15b) can then be used to convert all future levels to discharge so long as the channel at the site remains stable.

Building gauging structures

In cases where such rating equations are hard to compile, where high precision is required, or where river channel cross-sections are unstable, hydrologists build structures of metal or concrete across the stream. *Weirs* and *flumes* are the two main types. Weirs are rectangular or V-shaped barriers behind which water collects and then spills over (figure 16). The depth of spillage is related to discharge by a rating equation just as in the natural channel. With flumes the structure does not hold back a pool of water; the base of the flume is at the same level as the channel floor (figure 17). However, the walls of the flume are accurately curved to produce an invariable rating equation between depth of water in the flume and discharge.

Recording water level

To get our continuous record of river discharge we need to record river level with a float. This is generally housed to one side of the channel section, weir or flume, in a *stilling well*, connected to the channel by a horizontal *tapping pipe* (figure 15c). Thus, water level in the well is the same as water level in the river. The float is free to move up and down the well with the water level and its movements are transferred to a pen which marks a clock-driven chart (figure 15a). The whole assembly is usually housed in a hut for protection.

In the computer era, methods of recording water levels *digitally*, rather than on charts, have made data-processing easier. These instruments either punch holes into a paper tape or electronically record on magnetic recording tape; however, most of these still rely on the float and well arrangement.

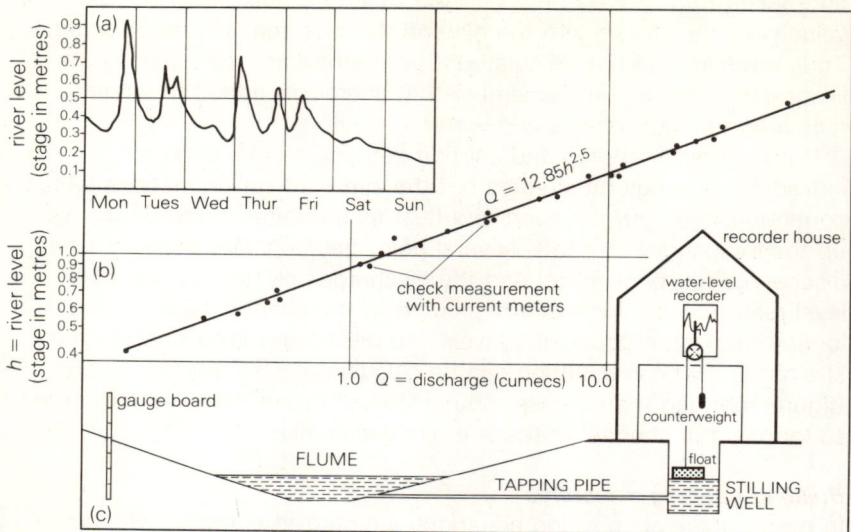

Figure 15 (a) Water-level recorder chart (b) Rating curve and equation
(c) Recording mechanisms

Measuring smaller flows

Not all flows of water in the catchment occur in recognisable
channels. Surface runoff or throughflow may need to be channelled
artificially in order to make a measurement. For surface runoff, roof
troughing carefully installed across the slope can be used to catch it.
Troughing can also be used to catch the throughflow from various
horizons of the soil if a pit is first dug carefully (figure 18). If the
throughflow occurs in cracks or burrows, down-pipe (also from the
builder's yard) may be used to lead it to the measurement site.

Small flows are much easier to measure than large ones. For a
point flow measurement of surface runoff, throughflow, or even the
discharge of very small streams during droughts, one can collect the
flow in a container of known volume (for example, a bucket or
measuring-cylinder) and time the duration of filling with a stop watch
or accurate wrist watch. This is the *volumetric* method. It can be used
to set up a rating equation or, if performed at close enough time
intervals, can be used to measure one site almost continuously;
surface runoff, for example may only last for a few hours during a
storm – but you must be prepared to get wet.

Figure 16 Rectangular weir being surveyed to establish true depth of water over the weir. This is, thereafter, recorded continuously by the recorder (in the box upstream) and converted to flow by means of the rating equation.

Figure 17 Trapezoidal flume. Observe the contrast with figure 13 – there is no pooling behind a flume.

Continuous measurements of small flows can be made with larger versions of the *tipping-buckets* used in rainfall recorders. Buckets of up to one-litre capacity work well. If an electronic system of recording cannot be arranged, a mechanical counter (like a bicycle mileometer) can be fixed to the device, being advanced one unit each tip by a cam on the buckets (see figure 19b).

Very small V-notch weirs are also usable with small flows; they are fixed to the wall of a weir-box which stills the flow and leaves enough room for a float to measure level.

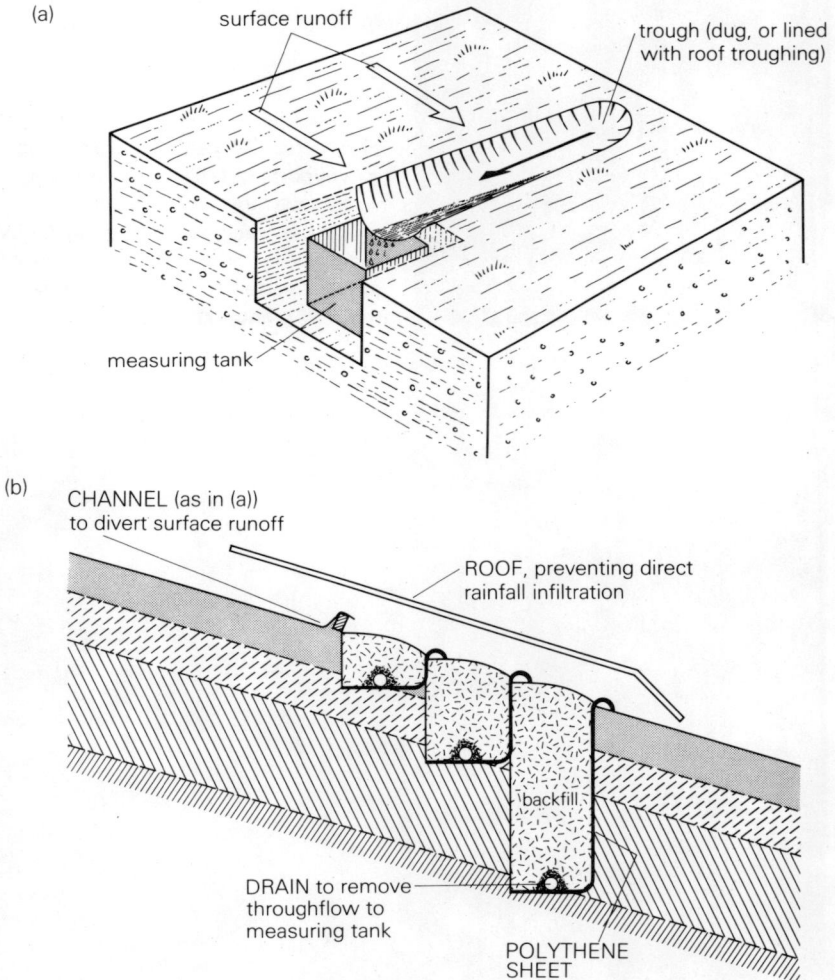

Figure 18 Simple suggestions for measuring (a) surface runoff or (b) throughflow, from three soil layers

Figure 19 Tipping bucket mechanisms in use in measuring (a) rainfall
(b) flow in a small ditch

Measuring soil moisture, groundwater and evapotranspiration

While losses are more often estimated than storages in catchment water-balance investigations, storage in the soil is the earlier process of the two during a rainstorm. Thus its measurement is treated here first.

Soil moisture

Although it shows a great deal of variation, an average soil is composed of 45 per cent minerals, 25 per cent water, 20 per cent air and 10 per cent organic matter. The water fraction is in various forms, *hygroscopic* (bound to soil particles), *capillary* (held between the particles against the force of gravity), and *gravitational* (free to drain from the soil and therefore mobile). The hydrologist is mainly interested in the mobile water, occupying the intergranular network of voids in the soil (see figure 20). This water plays a large part in *infiltration* and *drainage.* Capillary water plays a part in plant uptake and evapotranspiration. Obviously, physical soil properties such as the *texture* have a basic control over how much rainfall enters the soil as infiltration during a storm. They also control, in combination with climate, physiography and vegetation, the amount of moisture stored, drained or evaporated in the period between storms. Drainage (either laterally downslope or as deep *percolation*) is the main process between storms in winter, the upward movement to plants being much greater in summer.

Soil moisture can vary between two extreme states, *saturation*, in which all pores are water-filled, and *desiccation,* in which none is, but two more important states are recognised in the field, *field capacity*, the maximum amount of water that can be held at the end of free drainage and *permanent wilting point*, at which plants cannot overcome the suction of the moisture films round the soil particles. This suction or tension can be measured, and gives an index of soil moisture values. However, the weight or volume of moisture can also be assessed more directly.

Gravimetric determinations

If a known weight of soil (about 5 g), sampled with a *soil auger* in the field, is carried carefully in an airtight seal of foil to the laboratory and dried artificially (in an oven overnight at 105°C) the dry weight, subtracted from the original weight gives the weight of water driven off. This is normally expressed as a percentage of the original weight.

Two major problems exist with this method: it samples only a small fraction of the soil (soil moisture is very variable spatially) and it is a *destructive sample*. One cannot return to make repeated samplings on the same small area without destroying the natural moisture regime and ruining the site!

Non-destructive sampling

For this reason, less direct methods, calibrated by gravimetric determinations, have been developed. The most recent is the use of small radioactive sources to irradiate the soil (in harmless amounts). The so-called *neutron probe* can be lowered to any depth in a soil profile in a permanent *access tube* of light alloy (figure 21). One can come back repeatedly to the same tube and record the neutron count. *Fast neutrons* emitted from the radioactive source in the probe are scattered and slowed by contact with water molecules. The resulting *slow neutrons* are detected and counted, their proportion being a function of soil moisture. One problem is that the method is unworkable close to the soil surface (yet this is the most active layer of soil for moisture changes); however *surface extension tubes* have been designed which allow the true surface layers to be irradiated below a false column of soil.

It is also possible over large land-masses, to use the background radioactivity of the earth, reduced proportionally by soil moisture. It can be detected and measured from aircraft and provides a rapid soil moisture assessment for large areas.

Another indirect method, calibrated gravimetrically, is *electrical resistance*, in which two electrodes are placed within porous blocks in the soil.

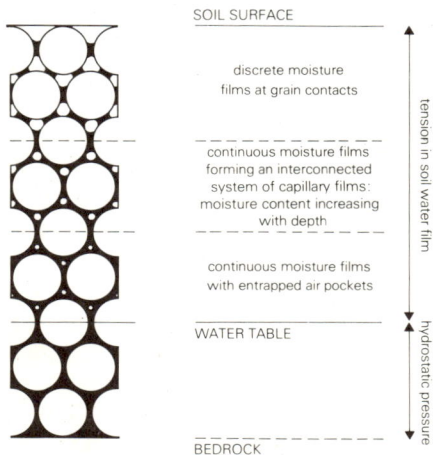

SOIL SURFACE

discrete moisture films at grain contacts

continuous moisture films forming an interconnected system of capillary films: moisture content increasing with depth

continuous moisture films with entrapped air pockets

tension in soil water film

WATER TABLE

hydrostatic pressure

BEDROCK

Figure 20   **Forms of soil moisture**

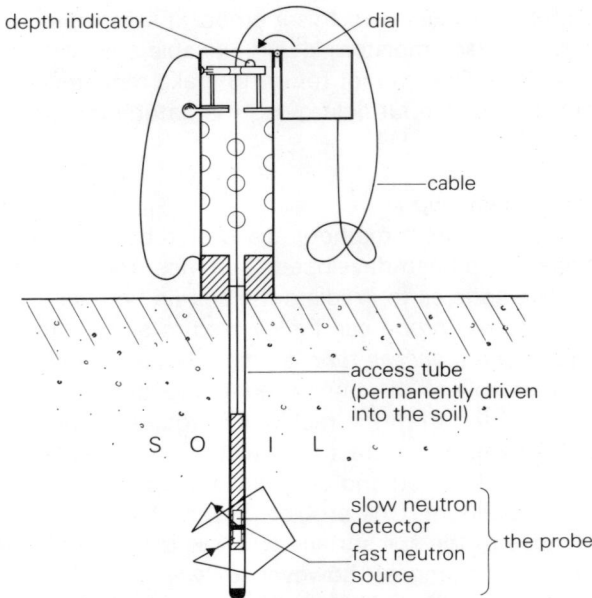

Figure 21 Soil moisture measurement using a 'neutron probe'. The probe is portable, and the readings easy, but the instrument is expensive, requires adequate safety provision, and the method requires calibration by more normal techniques.

Measuring soil water tension

Since water is acted on in the soil by various forces (gravity during drainage, capillary forces at field capacity), the variation of water content during infiltration, drainage, transpiration by plants, or evaporation from the surface, sets up tensions which can be measured. The *tensiometer* (figure 22a) consists of a water-filled porous cup, buried in the soil and linked by a continuous column of water (in a tube) to a manometer. Thus, as water passes from the cup into the surrounding soil by the tension set up during soil drying, this tension can be measured on the manometer. When the soil is wetting, the reverse process occurs. Since the tensions measured are in terms of a negative head of water of between 100 m and 1000 m, the unit chosen for tension units is their logarithm, or pF (just as pH is used for acidity). A low pF is wet, and a high pF is dry soil.

Obviously the relationship between tension and water content will vary between soils and the investigation into this *soil moisture characteristic* (figure 22b) tells us a great deal about the physics of soil moisture storage and movements. The tensiometer is most sensitive between saturation and field capacity – measuring at greater

(a)

scale against which
mercury level is read

mercury reservoir
nylon capillary tube

nylon tube
approximately 8 mm
bore

ceramic hollow
'candle' (porous pot)
approximately
60 mm × 15 mm

(b)

pF

sandy loam
- - - loam
········ clay–loam

soil moisture content (vol per cent)

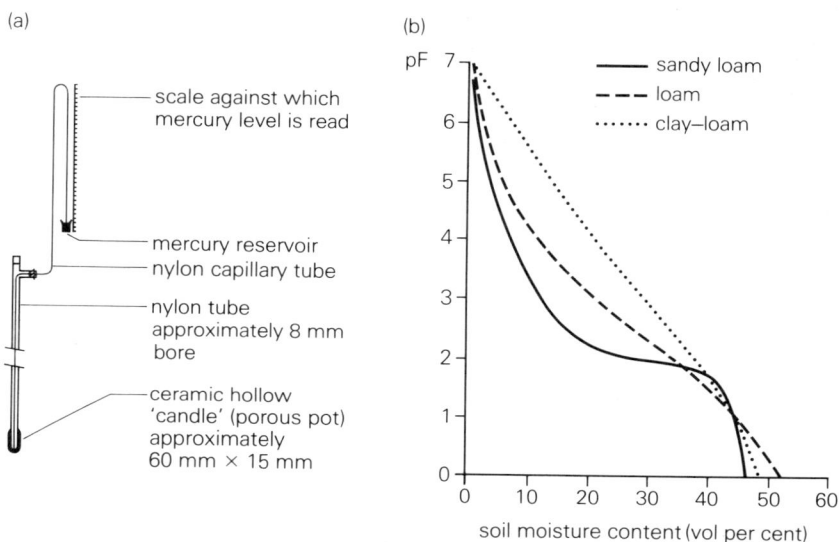

Figure 22 (a) A simple tensiometer (b) Soil characteristic curves.
*Copyright © 1975 McGraw-Hill Book Co. (UK) Ltd. From 'Principles of
Hydrology' 2nd edition by R. C. Ward. Reproduced by permission from
original diagrams.*

tensions is normally carried out experimentally in the laboratory using
pressure plate apparatus in which a vacuum is applied to the soil.

Infiltration

As well as measuring soil moisture in hydrology, we are also
interested in the rate at which it becomes wet during rainfall, in other
words the process of infiltration. Although simple techniques are
available to measure the *infiltration rate* of soils, variability in this rate
can be very complex and there are always doubts about extending our
measurements, usually made on very small plots, to whole catchment
areas.

A simple *infiltrometer* can be made from a tin can, strong enough
to be driven 30–50 mm into the soil. That part of the can still
projecting above the soil can be used as a reservoir for water added
by the student (figure 23a). The reservoir should be kept topped up to
a marked depth (say 10 mm) as its contents infiltrate the soil, rapidly
at first (if the soil is dry) and then more gradually. Keeping a note of
the quantities of water added over short time intervals will afterwards
allow the plotting of an infiltration rate curve (figure 23b). The steady
minimum rate, reached after most of the soil pores have been filled,

29

has value to engineers studying the runoff of heavy rainfalls to produce floods.

As well as investigating the vertical infiltration of rainfall into soils, hydrologists have also developed techniques for measuring the sideways movement of water in the soil, down slopes to stream channels (see pp. 22–25). Drainage of water from soil is vitally important to agriculture and artificial farm drainage has undergone marked expansion in recent years with the installation of the typical herringbone patterns of tile or plastic pipes beneath the cultivation layer.

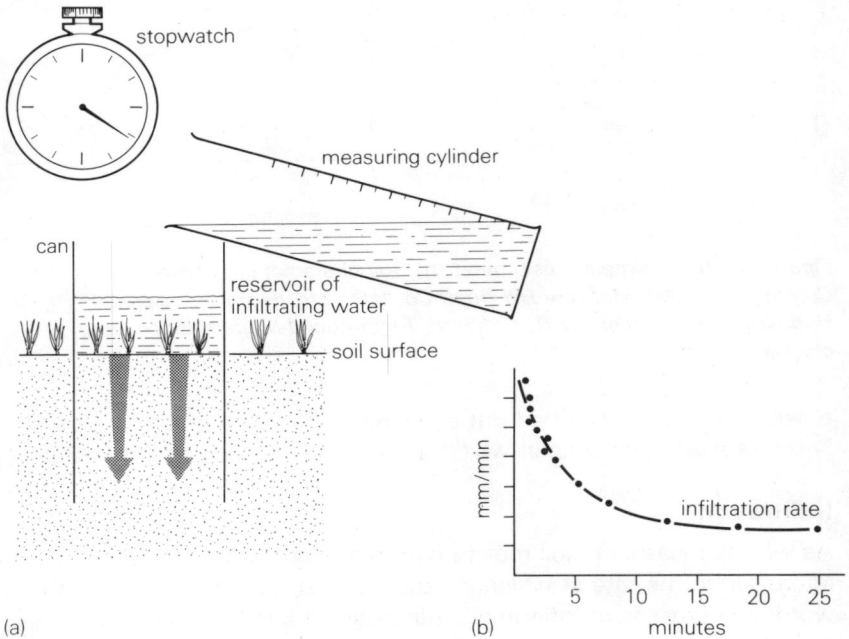

Figure 23 (a) An infiltrometer experiment. It is more efficient to replace the measuring cylinder with a 'constant head' reservoir which adds water automatically. (b) The results drawn up. Readings are taken rapidly at first, but as the rate slows and steadies, they are taken less frequently.

Groundwater

The fundamental phenomenon of groundwater is the *water table*. Between the soil and the water table in a permeable rock or aquifer is a *zone of aeration* through which water passes to reach the water table, also known as the *piezometric surface* since it may, in effect, be above ground in artesian systems. Water tables rise and fall according to the balance between *recharge* (from rainfall percolating downwards) and *discharge* from the aquifer in springs or direct to rivers. The property of aquifers is their *permeability*. Just as channel flow is powered by gravity but resisted by friction, the velocity of groundwater flow (*v*) represents the influence of the permeability (*K*) on the hydraulic gradient (slope of the water table = d*h*/d*l*). Thus *Darcy's Law* of saturated groundwater flow is:

$$v \approx K\frac{dh}{dl}$$

Velocities tend to be quite low in groundwater (10– 0·001 m/day in sandstones).

The major routine measurement made on our aquifers is that of water table depth. This measurement is either made down wells drilled to extract the groundwater or down specially-drilled boreholes. On impermeable rocks, with a good thickness of soil, the water table or piezometric surface may be measured in the same way, using a 100 mm diameter plastic down-pipe or tile-drainage pipe.

The instrument used for point measurements of water level is usually one in which the contact with water by a probe, lowered on a metal measuring tape, completes an electric circuit and rings a bell or lights a bulb. The sound or light equipment may be housed on the surface or in the probe itself (see figure 24).

Figure 24 'Diptone' probe for measuring water levels in soil or levels of groundwater

Continuous water table level measurements may be made using float and pen recorders not dissimilar to those used for river level recording. However, groundwater fluctuates much less rapidly than surface streams and hence less regular readings can be taken in many cases without loss of accuracy.

Tests are made of groundwater flow rates by pumping water from wells to cause *drawdown*. Either the fall of water levels in neighbouring test boreholes or recovery of the main well on cessation of pumping are observed. The slope of the drawdown at neighbouring wells at various rates of pumping or the rapidity of recovery can both be used to calculate permeability from Darcy's Law, because the velocity of flow and the water table gradient are both known (see Darcy's equation above).

Evapotranspiration

Although circulation in the hydrological cycle depends mainly on the flow of rivers to the oceans and evaporation from their surfaces, there is a major 'short circuit' in evaporation from land areas. This comprises evaporation of small storages in depressions, on leaf surfaces etc., and transpiration by plants from the soil store. Normally the two processes are considered jointly under the term *evapotranspiration*. Transpiration is the dominant control over loss of water from most temperate land surfaces and therefore direct measurement is difficult since the plant/soil/atmosphere system cannot be interfered with too much with instruments. Therefore, important indirect methods have been developed.

Direct measurement

Consider the puddles which remain after heavy rain. It is possible to observe them gradually reducing in size through evaporation; sometimes in summer one can observe wet surfaces such as pavements, roofs and trees steaming. The simplest direct measurement of this *open water evaporation* is to set up an *evaporation pan* (figure 25). Two main types of metal pan are used, the circular (1·22 m diameter, 0·25 m deep) United States version, which is set above the ground and the British square pan (1·83 m x 1·83 m x 0·61 m deep), which is set in the ground. Rainfall is measured nearby (the pan is normally on a meteorological site) and, if not sufficient to maintain water level in the pan, known amounts are added to keep the water level constant (or the fall of water level measured on an accurate dial). Refinements of the evaporation-pan principle seek to imitate the action of soil and plant cover in controlling the loss of moisture. The *atmometer* consists of a porous membrane across a store of water; the *evaporimeter* has a cork-filled column above the water surface of a raingauge. The fullest expression of simulating the true plant/soil/atmosphere relationship is the isolation of a column of soil and vegetation in a container which is then reburied in the site from which it was taken, with the minimum of disturbance. This is the principle of the *lysimeter*. Lysimeters are of various sizes, the larger being generally the better, in that disturbances along the edge have a reduced proportional influence on the main block of soil and vegetation. The largest lysimeters are 'natural' – no soil is moved but the plot is isolated by lateral trenches and metal plates. At the smallest scale a gallon can forms a weighable tank lysimeter (figure 26). In the operation of a lysimeter one is attempting to measure the water balance of a small model catchment. Thus rainfall measurements must be taken nearby and percolation water collected from the base of the soil column. Surface runoff must also be collected if this occurs.

Figure 25 (a) United States type of evaporation pan (b) British type of evaporation pan

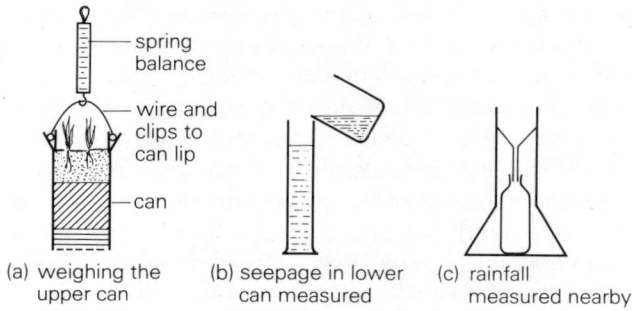

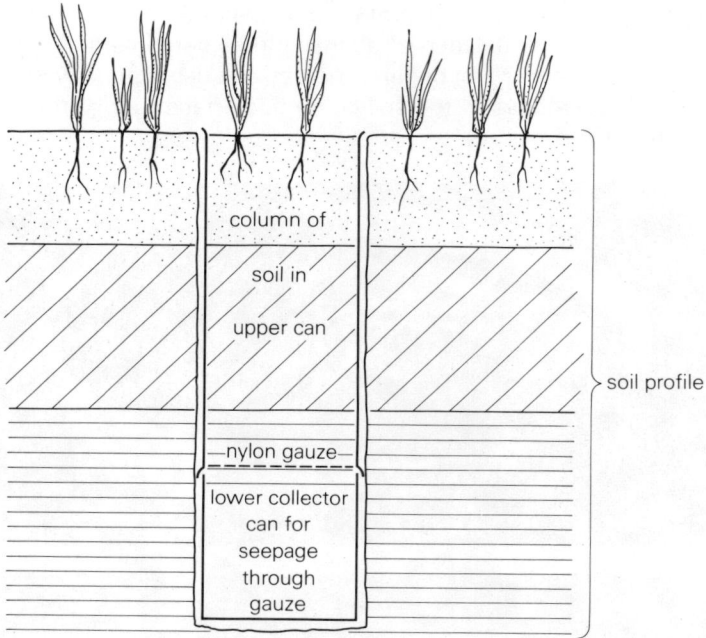

Figure 26 **A simple weighing lysimeter constructed from food or drink cans. Routine measurements are shown in (a), (b) and (c).**

The changes of storage in the soil column can be established by indirect soil moisture measurements (clearly, removing soil samples is not allowed from a lysimeter) or by weighing the whole column (in the case of small tank lysimeters).

Thus, if at the end of a week, a small 200 mm diameter lysimeter weighs 1 kg more than it did the week before, there has been 50 mm of rainfall and 10 mm of seepage – the missing quantity must have been lost by evapotranspiration. Work it out; you only need to know in addition the value of π and the fact that 1 cm³ of water weighs 1 g.

Indirect measurement

The main meteorological processes controlling evaporation are basically the humidity of the air at the prevailing temperature, the radiant energy of the sun and the movement of the air.

The early methods of expressing evaporation as a function of temperature (such as Thornthwaite's) have been used for climatic estimations. However, for hydrology a method combining all the major factors was developed by H. L. Penman. Meteorological observations of radiation (or sunshine), wet and dry-bulb temperature and windspeed are fed into the Penman equation and the effect of vegetation differences can be included by their *albedo* (their reflectivity to radiant energy).

The only effect not included is that of regulation of the evapotranspiration by the plant itself. The plant's rate of water use depends on its being capable of withdrawing water against the tensions in the soil. Penman's *potential evapotranspiration* often, therefore, overestimates *actual* evapotranspiration and consequently he included the term *root constant*, a soil moisture content below which the actual rate falls below potential. Obviously deeper-rooted plants can tap water supplies at suitably low tension long after shallow-rooted ones have cut down their intake.

Potential evapotranspiration calculations can also underestimate actual losses from vegetation, when a substantial proportion of the loss occurs as evaporation of rainfall intercepted on the vegetation canopy, for example losses from coniferous forests in upland areas (see pages 53–54).

Soil moisture deficit

The Penman calculation is performed regularly on climatic data at 176 sites in Great Britain to give an estimate of evapotranspiration. Since rainfall data are also available, the main input and loss to the land phase of the hydrological cycle can be interpolated for the whole of Great Britain. The Meteorological Office run a kind of bank account, in which deposits of rain are matched against withdrawals by evapotranspiration (runoff is only considered to occur when rainfall exceeds evapotranspiration, that is, for most of the winter).

Fortnightly bulletins on the state of the account are issued during spring, summer and autumn when accumulated evapotranspiration begins to exceed rainfall. The water balance is then 'in the red' and the number of millimetres of rain needed to bring it back to field capacity (see figure 27) are mapped as the *soil moisture deficit (SMD)*. Farmers and growers use the figures to guide them in irrigating crops – there is clearly no point in adding water beyond what is needed. Water authorities use the figures to base predictions of flooding, subtracting the soil moisture deficit from the forecast rainfall to give an estimate of surplus rainfall, hence runoff to rivers.

It is important, therefore, to remember that soil moisture *measurements* are not yet performed as a matter of routine in the United Kingdom, except as part of research. The regular estimate of soil moisture is an indirect one via rainfall and evapotranspiration. Sampling the wide variability of soil moisture by direct measurement would be very expensive and time-consuming; in any case, at the national or regional scale, only a broad guide is required, and this is best provided by SMD.

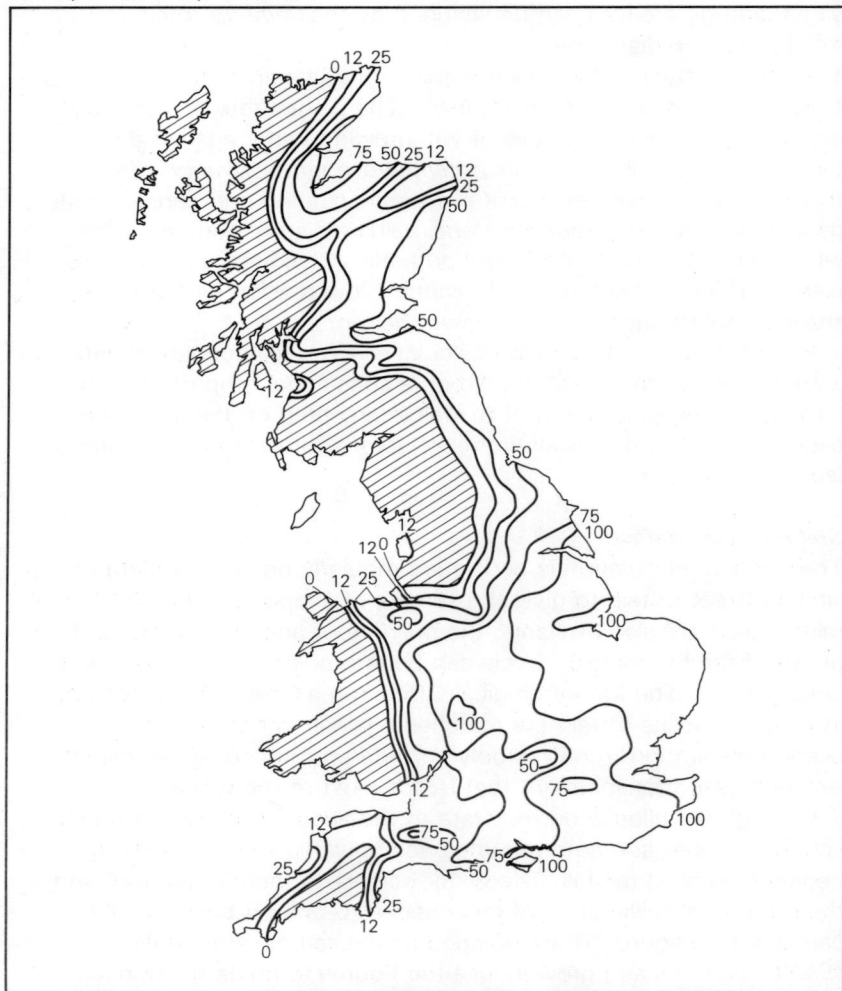

Figure 27 Soil Moisture Deficit. Estimated SMD at 0900 GMT 27 September 1967, over Great Britain.

Extremes – water: hazard or resource?

In this chapter we turn from measurement to application, from hydrological research to hydrological practice. While geographers, physicists, biologists and mathematicians all contribute to the study of hydrology as a science, in the applied field the civil engineer dominates. He holds this place because his is the longest background in history: scientists are newcomers by comparison. For centuries the engineer has had to produce practical solutions to the problem of too much or too little water; he has also borne the responsibility for any failure in his schemes.

Floods

The small town of Lynmouth in North Devon was devastated on the night of Friday, 15 August 1952 by a flood which exhibited all the features of this major natural hazard, concentrated into a few fatal hours.

The riverside location was a picturesque one near the confluence of the east and west branches of the river Lynn and also the sea. Rows of cottages and shops crowded the constricted banks of the river; a narrow, rustic bridge crossed from one bank to another. Upstream in the steep-sided gorges of the Lynn, the forested slopes plunged into the boulder-strewn channel with trees hanging over the water. On Exmoor's boggy soils, near-saturation had been reached in the preceding days by falls of rain and drizzle. On the better soils, agricultural drainage had produced an improvement in subsoil runoff. Onto this saturated catchment, a rather localised storm added 225 mm of rain for the 24 hours up to 0900 GMT on the 16 August when raingauges were read. Most of the rain fell in two intense spells at around 1700 GMT and 2100 GMT on the 15 August. The Lynn rose astonishingly quickly and ripped down the gorges from the moor to the town, surging as successive flood waves were released from temporary dams behind the boulders and trees littering the channel. In the town the capacity of the confined channel and bridge was quickly exceeded and houses were swept away as the torrent broke its banks. Thirty-four people died, 90 homes were destroyed, 130 cars and 19 boats were swept out to sea. The total damage cost £9 million, while the redesign of the channel through the town after the flood had to be based on estimates of rainfall and runoff since no adequate measurements were available.

Causes of floods

Floods as a hazard to man are chiefly the result of his attraction to water. Most settlements, among them the major ones of the world, adopt the river bank or floodplain site and consequently risk flooding. The physical cause is, of course, abnormal precipitation, either of much longer *duration* or of much greater *intensity* than normal. Perhaps the same amount of rainfall is involved in each type of flood. Figure 28a shows that rainfall is basically either 'short and sharp' or 'long and steady'. The former type of storm is most calamitous in small steep catchments such as the Lynn whilst the latter tends to be worst in large catchments. One must also appreciate that the short, sharp rainstorm is normally also limited in territorial extent whereas the longer, steady storm may cover the whole country (figure 28b). Typically the former is caused by *convection* as in a summer thunderstorm, and the latter by *frontal conditions*, these types occurring dominantly in summer and winter respectively. Figure 28a shows that, as the duration of a storm increases, the average intensity of rainfall for the storm decreases. Similarly, figure 28b shows that very deep falls of rain seldom occur over large areas; as the area of a storm increases, therefore, its average rainfall over the area decreases.

Melting snow can also cause flooding. In many countries of the north, each spring brings snow-melt floods down from the mountains. In Britain, rainfall falling on melting snow produced the widespread flooding of 1947 and many floods are exacerbated to some extent each year by the addition of melting snow to rainfall. However, remember that it takes nearly 8 mm of average snowfall to produce 1 mm of rain equivalent.

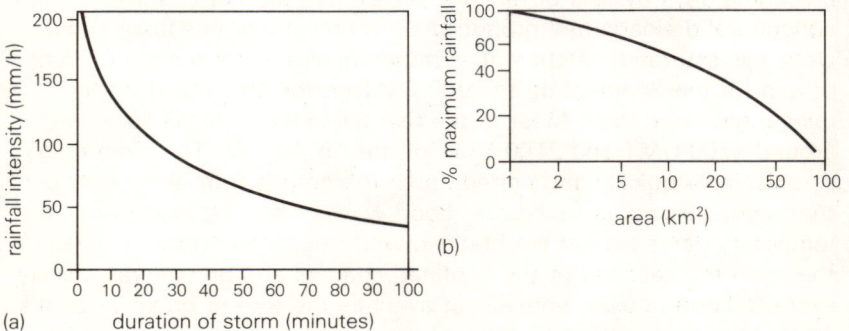

(a) duration of storm (minutes)

(b) area (km^2)

Figure 28 (a) Rainfall intensity/duration (b) Rainfall/area relationship. Principles of flood-producing rainstorms – 'the heavier the shorter, and the heavier the smaller'.

Measurements during and after floods

To be able to understand we must measure. Yet floods upset most of our hydrological instrument systems. Often they occur in catchments without instruments, as was the case at Lynmouth. The raingauge, even if one is present, may be filled to overflowing, swept away if near the river, or buried by a landslip. The rainfall recorder may stick as the float and pen system begin to rise and fall at great speed during an intense storm. The river-level recorder may also malfunction; although it is more likely that our gauge-board or even the weir or flume will be destroyed or damaged: even if they are not they will almost certainly have been bypassed as the flood flows round the measurement station on each side. It is possible to attempt gauging with a current-meter from a safe bridge in such conditions, although the meter must be weighted enough to hold it steady.

The most practical record to make during a flood is a photographic one. Photographs are worth a great deal of eye witness accounts (see figure 29). Obviously instrumental records of rainfall and river level are the most valuable but can be backed up or even substituted with estimates from the rain collected in a wheelbarrow, milk churn, animal feed-trough or similar containers, or from the extent of the flood marks left by the swollen river. To derive a rainfall estimate the 'flood detective' must measure the collecting area of the makeshift vessel concerned and calibrate its catch accordingly. Using the flood marks (dead leaves, gravel, hay, litter, etc.) by the river one can calculate the cross-sectional area of the stream at the height of the flood. The degree of slope downstream of these marks then gives the other component of discharge velocity by using an equation bearing the name of R. Manning. *Manning's 'n'*, the figure which accounts for frictional resistance to flow ('roughness'), is around $0 \cdot 030$ for lowland rivers and $0 \cdot 045$ for rocky upland channels.

$$\text{Peak discharge } (Q) = \frac{(A \times R^{\frac{2}{3}} \times S^{\frac{1}{2}})}{n}$$

A = cross-sectional area of channel at peak flow

R = hydraulic radius of channel at peak flow: $\dfrac{A}{w + 2d}$ where w = channel width, d = channel depth.

S = slope (as a decimal)

n = Manning's n.

Remedies – prediction or forecast?

With good flood data (for both rainfall and runoff) there are now many sophisticated methods of analysis available to the hydrologist. They may be grouped into two classes based on what they aim to do.

Figure 29 The Newtown floods of 1964. These photographs allow quite detailed estimates of the very high river flow.

The first class, most common to date, involves using long records of floods over the years to calculate the average *recurrence interval* of a certain sized flood. This is the *magnitude-frequency* approach and is normally complemented with the *unit hydrograph* approach for extreme floods which analyses the timing of the rise and fall of flood waters. These two methods both *predict* what the features of a flood in the future will be so that we can build river walls, bridges, dams and even simple culvert drains to an adequate size and strength to protect the community. Alternatively, we can leave the defences down and issue warnings to people to move out in times of flood – we *forecast* the flood's height and timing based on our knowledge of the rainfall imminent and the behaviour of the runoff process over the catchment.

Dealing with *prediction and protection* first, the first requirement for any statistical analysis of floods is a long record of data for the flow of the river in question or a good *correlation* of its short record with the appropriate part of a longer record from a catchment in the same region. More commonly, only the *annual peak flow* is extracted from the lengthy record and these annual peaks are fitted to an *extreme value distribution*. By using Gumbel/Powell graph paper (figure 30) and a plotting formula, the annual peaks can be fitted to a straight line and the recurrence interval (or *return period*) can be read off on the x-axis for any flood on the y-axis of the graph. In practice considerable statistical expertise is required in using such an analysis.

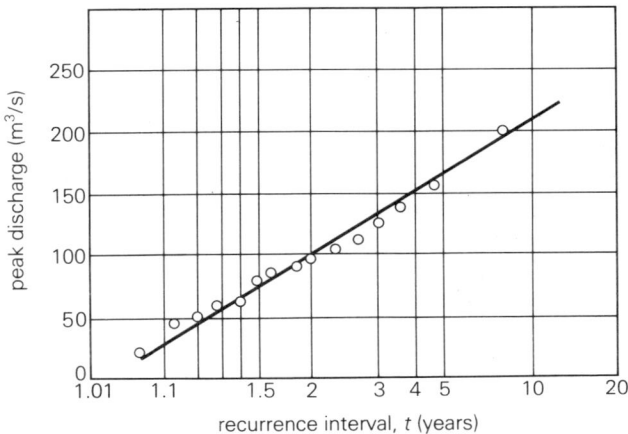

Figure 30 Flood frequency analysis, using Gumbel/Powell plotting paper. The points represent the annual peak flows during the streamflow records, plotted after ranking by size, according to the formula $t = \dfrac{n+1}{m}$ where n is the number of years on record, and m is the rank of the flood.

The engineer will normally build a protection scheme based on the estimated peak discharge of, say, a 100-year return period (for towns threatened by major rivers) 10 years (for storm sewers) and so on. The return period is obviously the average interval between floods of a certain magnitude; frequently the public take it to mean that they are safe from floods for 10 years or 100 years but the 10 or 100 year flood may always occur today, or even be reached in two successive years!

The hydrograph approach to floods analyses the relationship between the catchment and successive hourly increments of rainfall and river flow (figure 31). Commonly the hydrograph is thought of as a triangle whose basic dimensions are *time-to-peak*, *peak rate of discharge* and *time base* (figure 32). The unit hydrograph is so-called because it represents the runoff from a unit of rainfall, say 25 mm, as a means of standardising over many floods and many catchments. These hydrographs can then be used with estimates of very high return period rainfall (rainfall records being much longer for statistical analysis than stream flow) to predict very high return-period floods. Such predictions are made for the design of dams since no dam should be overwhelmed during its economic lifetime – the consequences would be too serious. Some dams are in fact now designed and operated specifically to contain all floods, for example, the Clywedog Dam in mid-Wales (figure 33); this makes economic sense in all respects – stopping damage by flooding and storing valuable water resources which would otherwise be 'wasted' in the flood and possibly cause damage.

Flood forecasting is as yet only possible on the bigger rivers, down which the flood wave takes days to move. On the shorter, steeper streams which are typical of the British Isles, protection, not warning, is generally the rule. However, advances in rainfall forecasting, matched with the unit hydrograph or a mathematical catchment runoff model may mean better river control in all respects in the coming years. Nevertheless, better planning of land-use in relation to river flood plains will also aid reductions in flood damage, especially around towns. It cannot be too strongly emphasised that floods are a natural phenomenon and no conceivable system of protection could control them completely.

Droughts

If you attempt the calculation of how big a collector for rainfall you would need to supply yourself with water, given average daily rainfall (see page 55), how would you cope with ensuring your supply for 15–30 rainless days? This is the problem faced by the water supply engineer on a national scale. Droughts are defined for the United

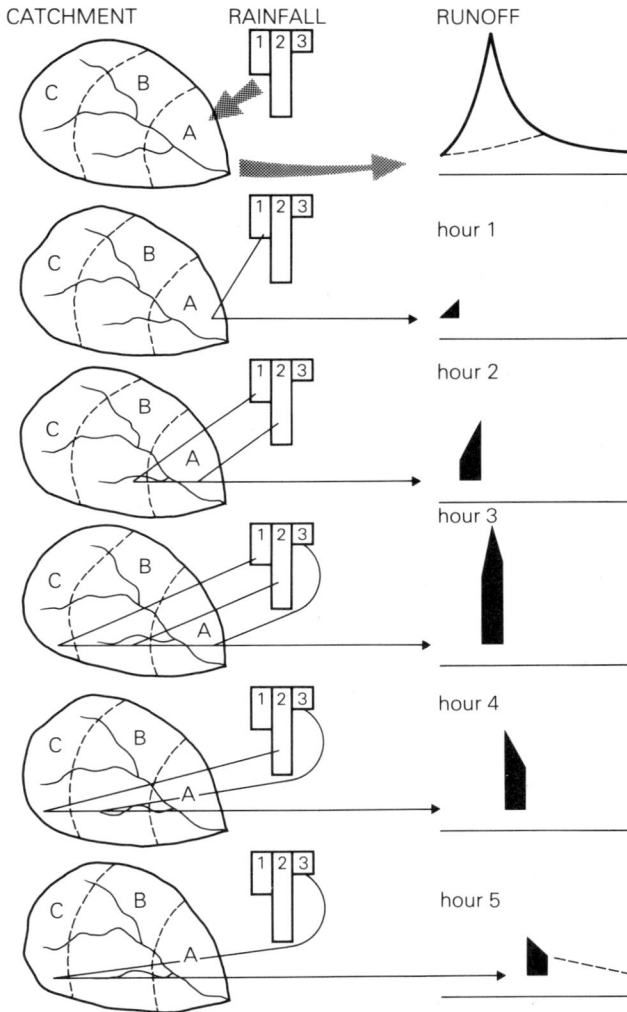

Figure 31 The principle of the streamflow hydrograph. The relationship of three hours of rainfall (downward-pointing bar-graphs) with three zones of a catchment (left) to give successive hours of runoff (right). In hour one, only the rain in hour one falling over zone A reaches our river gauge. By hour two, the rain in hour two over zone A is joined by that rain from hour one which fell further away in zone B . . . and so on.

Kingdom as *absolute drought* (15 days with less than 0·2 mm of rain on any day) or *partial drought* (29 days with an *average* daily rainfall of less than 0·2 mm).

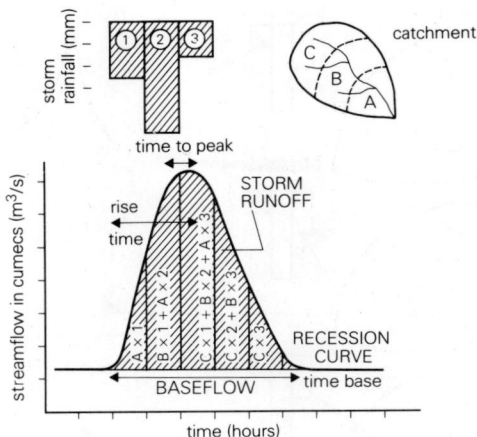

Figure 32 Summary of principles and terms relating to the flood hydrograph

Causes of drought

The normal distribution of pressure systems and air masses favours the British Isles with a well-distributed annual rainfall from the Atlantic. However, there is a commonly-recurring pressure pattern which often results in *'blocking' anticyclones* whose strength and persistence wards off the prevailing westerlies and frontal systems. An easterly or southerly airstream over Britain has had only a limited oceanic passage in which to pick up water vapour for precipitation. Droughts are therefore most frequent in the southern parts of England and East Anglia.

Normally rivers and aquifers subjected to a dry summer in the British Isles are recharged by winter precipitation, winter being a period in which the westerly circulation normally reasserts itself. The most serious droughts therefore result from two consecutive dry summers with an intervening dry winter (for example 1975–6 where the April–March twelve-month period was the driest since rainfall records began over England and Wales, see figure 34). Even then, Scotland and the North fared better than the South.

Water supply

The answer to the question at the beginning of the section on drought on supplying oneself with water from rainfall when there is none is, of course, to provide *storage* when there is rainfall. We already know that storage is a natural part of the hydrological cycle – for example in the soil moisture and groundwater systems which sustain the flow of rivers long into a drought. This was the attraction of river bank or springline settlements to our ancestors. It is also the

Figure 33 The Clywedog Dam, mid-Wales. This photograph illustrates the problems of operating a major storage of water. In 1974, so much rain fell that no more could be stored, and the excess had to be allowed to overflow. The dam is designed to be safe even under such conditions.

basis of water-supply which has a well as its main source.

The responsibility of providing extra storage for our own supply, for industry and agriculture is now taken over by water resource engineers on a regional or even a national basis.

They rely heavily on our *average* figure for rainfall, established over a minimum of thirty years of record. Yet rainfall varies from year to year (figure 35) and in an extreme year may be as much as 150 per cent of average or as little as 60 per cent. Reservoirs are built to maintain supplies for the three driest years in a hundred, a period over which only 80 per cent average rainfall is expected.

River flow analyses are also used in the study of droughts, although as in the case of floods, they are of shorter duration than rainfall records and therefore less value for very rare extremes. Just as in the analysis of the return period of floods the annual peaks were plotted on probability paper (figure 30), for droughts the lowest flow reached each year is plotted. Engineers also study the recession curve of the river as its level gradually falls in order to be able to predict the date at which, if no more rain falls, the river will run dry or water resources will become critical.

Figure 34 Drought. The Clywedog Reservoir in Summer 1975 – the other end of the hydrological extreme, showing equipment from one of the farms submerged when the dam was constructed.

Public water supply as we know it today did not begin in Britain until public health improvements necessitated reservoir construction for the cities of northern England during the second half of the nineteenth century.

Since storage is of little use without distribution, most of the older reservoirs are linked by pipeline to a major city or town. However, this has traditionally restricted other public uses of the reservoir such as swimming, boating and fishing because pure water is cheapest to process. Pipelines are also costly to maintain. Consequently the most recent reservoirs have been built to *regulate* natural rivers, smoothing out the natural extremes of floods and droughts to give a constant supply to any town or city on its banks. The Severn, Dee and Tees are already regulated and there are plans to link up regulation schemes to other rivers to create a *water grid* rather similar to that used for electricity. Another modern approach to storage for droughts is to use the groundwater reservoir below our feet. This 'reservoir' does not need a dam to be built, only a means of *artificial recharge* to fill it. This has, however, proved to be quite a difficult technique. Much has been written of *estuarine storage* and *desalination* as water resource schemes but these are unlikely to replace reservoirs before the end of this century; indeed, as this book is being written, approval has been given to construct the biggest reservoir in Europe in mid-Wales (Craig Goch). This will regulate the Wye, possibly the Severn and indirectly the Thames. (See figure 36.)

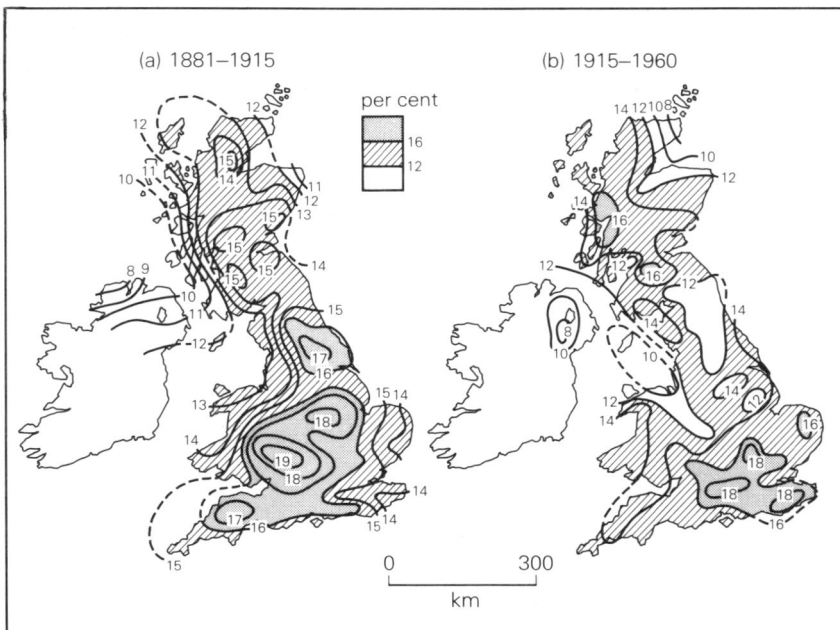

Figure 35 Percentage variation of annual precipitation in the United Kingdom. It is essential to consider the variability as well as the mean in water-resource planning. Obviously, the reference period taken is also important from a consideration of the differences in these two maps.
Copyright © 1975 McGraw-Hill Book Co. (UK) Ltd. From 'Principles of Hydrology' 2nd edition by R. C. Ward. Reproduced by permission from original diagrams.

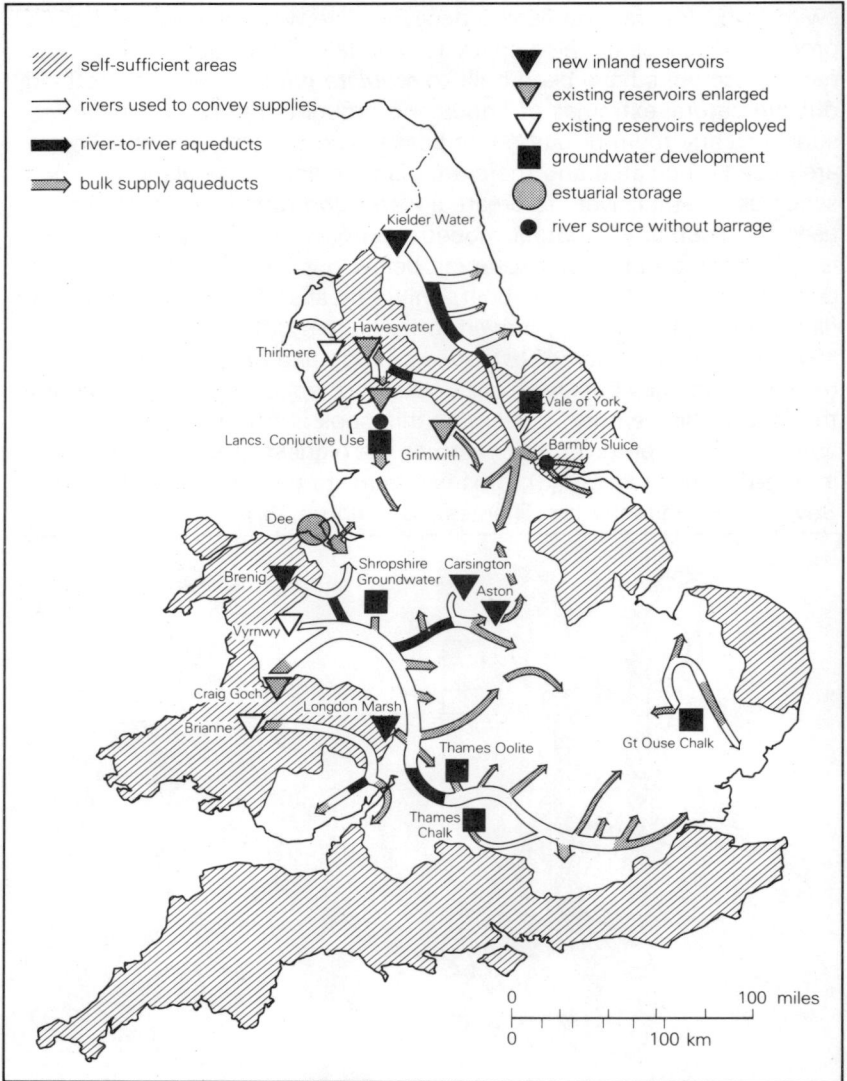

Figure 36 A national strategy for the development of a new water supply capacity

Agriculture

The first storage to run out during a drought is the interception store on the vegetation canopy (this often happens in a matter of hours). Plants then begin to transpire and evapotranspiration next exhausts the soil moisture store from the surface down. The rooting properties of various crops and their tolerance to drought determines how much

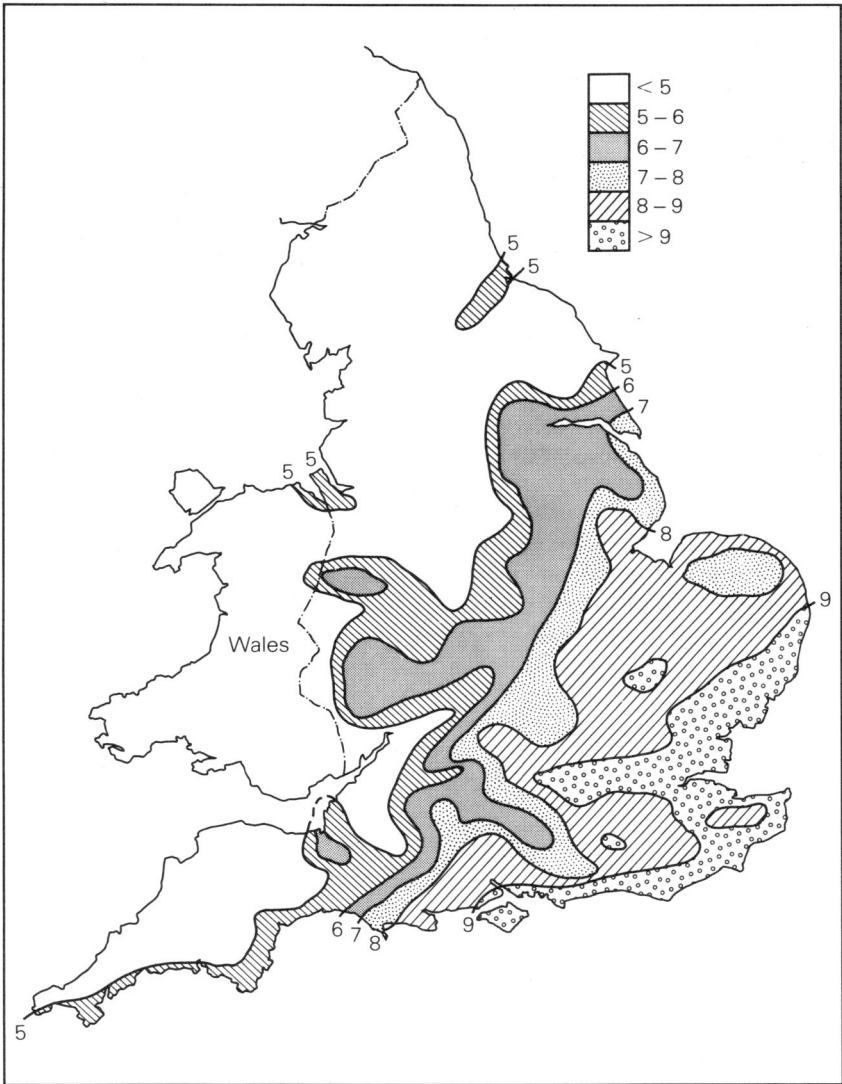

Figure 37 Frequency of irrigation need (years in every ten)

water the farmer needs to add by *irrigation* and also when to add it. He mainly works from the Soil Moisture Deficit calculated by the Meteorological Office (see pages 35–36). Most farm land east of a line from the Exe to the Humber in England would benefit from irrigation in more than five years in every ten (figure 37) but this is precisely the area where water is in shortest supply and most demand. Farm reservoirs or boreholes are often the answer.

The future of predictions and forecasts in hydrology

One of the aims of any science is to give insight to the future by collecting observations on the present or interpreting the past. Although hydrology is a young science it is already in a position to begin helping the engineer solve our flood and drought problems. During the first long phase of practising hydrology, engineers had to rely on their experience and cover themselves against error by building bigger dams, higher banks and wider bridges than were really necessary in economic terms. After hydrologists had collected enough data, the estimates of floods and droughts could be improved and matched up to the *costs* of developing protection and the *benefits* of doing so.

In the next phase an understanding of all the complex processes which result in a flood or drought may be used to develop mathematical models of river basins. These models, linked to instrumental improvements – measuring rainfall by radar, satellite weather-forecasting and so on – will no doubt end in large river basins being controlled from one centre, manned by a hydrologist or even automatically by computer. However, this phase, even if completely successful, will not mark the end of scientific hydrology's aid to the engineer. Natural systems are constantly changing; certain changes may be brought about by Man himself and these are the topic of the next chapter.

The influence of Man . . .

. . . on climate

Climate is constantly changing: the records of the past, both instrumental and geological, tell us that it is seldom stable for long. Cycles of various periodicities are superimposed upon each other to a degree which makes the detection of reliable trends very difficult. However, one set of trends can be discerned that parallels the impact of man and his technology on the global and local behaviour of the elements. Some of man's influence is inadvertent, some is purposeful. Under the first heading comes atmospheric pollution, under the second weather-modification programmes.

Since the power for the global atmospheric engine (and hence the global hydrological cycle) is solar radiation, changes in the atmospheric envelope of the globe clearly have a direct, if slow, influence on the cycle. For example, the carbon dioxide concentration of the atmosphere has increased by 10 per cent this century and is rising at $0 \cdot 5$ per cent per annum. It absorbs the long-wave earth radiation and has a heating effect. However, atmospheric temperatures have in fact fallen, due to the other influence of Man – the addition of dust and aerosols to the atmosphere (these intercept short-wave solar radiation). There are allegations that air pollution by the heavily industrialised areas of the northern hemisphere have, through affecting circulation patterns, caused drought in the Sahel zone of Africa.

The effects of individual cities on rainfall are beginning to be assessed. London's increased rainfall from thundery outbreaks is now thought to be as much due to the physical obstruction of the city to atmospheric movements as to the extra warmth, moisture and raindrop nuclei produced by its people and industry (see figure 38). The Meteorological Office has predicted that atmospheric warming by the big nuclear power plants of the future may lead to increased rainfall up to 1000 km downwind. A likely hydrological effect of the irrigation of desert areas, which has been necessary to keep pace with the demand for food of the accelerating world population, is that evaporation over land areas will increase by 200 mm per annum and as a consequence world precipitation would increase by 6 per cent.

There have also been attempts to solve problems of water deficiency by experiments to augment rainfall or make it occur more evenly. 'Cloud seeding' has a lengthy history but so far can only encourage precipitation, not cause it.

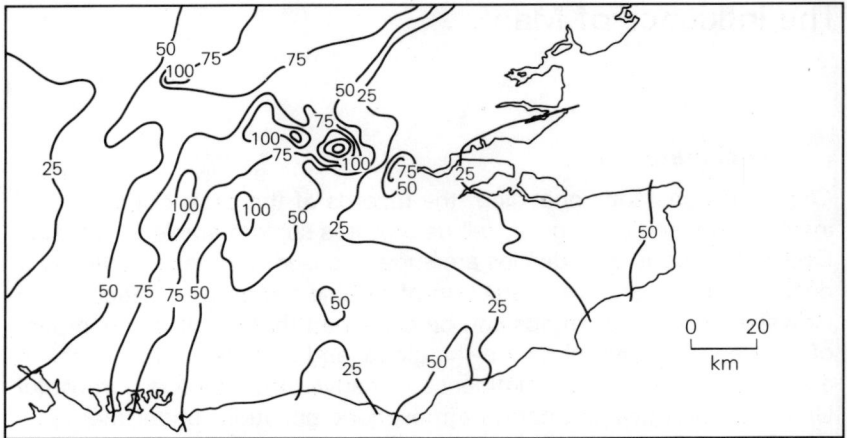

Figure 38 The influence of London on precipitation. Thunder rainfall (mm) in warm front situations, summer 1951–60.

. . . on runoff

The natural interface between rainfall and runoff is the soil and its vegetation cover, where present. All but the worst-drained, least permeable soils and all but the lowest plants provide storage capacity for precipitation. When these are removed or replaced by man, changes in the runoff process are bound to occur. In the interests of agriculture and settlement large areas of *forest* in the world have been cleared. Hydrologists in the United States pointed out in the first half of the century that such removal resulted in increases in runoff and erosion of soil on steep slopes. In the spread of concrete and tarmac surfaces during *urbanisation*, most of the natural storage capacity is replaced by an impervious surface and a new system of artificial drains installed to replace natural channels. It has been proved in one of Britain's new towns, Harlow, that the peak discharge from urbanised catchments is higher and the rate of rise faster for all moderate floods contained in the storm drains (see figure 39). For some large floods, too big for storm drains, the effect is not so pronounced because of storage provided in the form of flooded roads, basements and so on. Among devices developed in the United States for easing these effects on runoff, are porous pavements and roads and large storage tanks on flat roofs of houses.

The reduction of natural hydrological storage by Man also has an effect on drought – storage sustains flow during rainless periods. *Drainage* of all kinds, urban, agricultural and forestry, superimposes an artificial influence on the channel phase of runoff. However, one

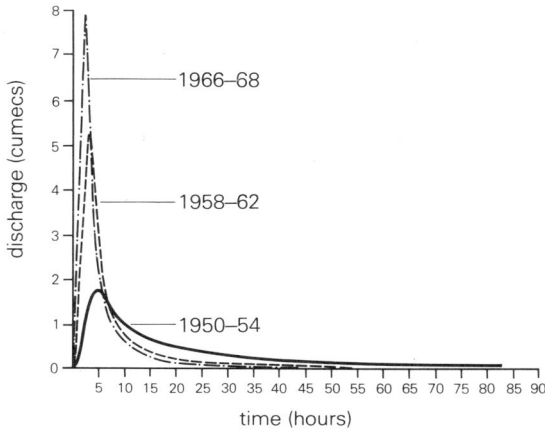

Figure 39 The influence of the growing urban cover of Harlow New Town on the hydrograph of Canon's Brook, 1950–68

should always be careful to consider both sides of the coin. Although artificial drainage may be seen to increase drainage density and hence flood peaks, one should bear in mind that, by reducing the moisture stored in the soil between storms, more storage is available for infiltration of rainfall during the next storm. Also, where drainage is the forerunner of a larger stand of vegetation (for example, drainage in the uplands prior to coniferous afforestation), that vegetation canopy also eventually provides extra storage. As with urbanisation there is also a corollary for periods of water shortage: a well-drained field may need irrigation during the summer months, the drainage having removed the store of soil moisture so unwelcome in the winter! Coniferous afforestation is accused of 'wasting water' by transpiring water from the soil during dry periods; however we now know that the interception and re-evaporation of rainfall from forest canopies (of all sorts) is the main effect.

One of the major international efforts to research the effects of land use on river flow has centred on the rivers Wye and Severn on the slopes of Plynlimon in mid-Wales. Here the runoff coefficient (see figure 23) for the grassland Wye is 82 per cent, but for the forested Severn it is 70 per cent. Not all the Severn is forested and the runoff coefficient for the trees is estimated at 62 per cent. The rainfall for the two areas is similar and the extra loss of water from the trees is put down to the evaporation of intercepted water from the dense canopy of coniferous trees. Obviously the effect of the tree crop changes during its lifetime – in the 1960s new planting (with deep drainage) in

53

the forests of the area was accused of increasing runoff rather than decreasing it. Thus, one can hypothesise the gradual amelioration of the drainage as water levels decline in the soil, giving storage and then the gradual closure of the tree canopy to give yet more storage and higher potential losses through evaporation.

Water resources, floods and droughts, are not yet controlled in this country by positive controls on land use in catchment areas. However, over much of the world the agriculturalist and town planner are as important in flood control as the engineer (for example in the United States and New Zealand).

. . . on water quality

Water is said to exist in three states: too much, too little or too dirty. Dirty water cannot be used for drinking, agriculture or industry without very expensive purification procedures. The best argument is that this expense should be concentrated on the premises causing the pollution in the form of investment by the polluters. The Control of Pollution Act has now given legal backing to a cleanup of British rivers which has already made good progress (freshwater fish are now common again in the Thames). The obvious industrial polluter, emitting coloured dyes, poisons or hot water from its cooling plants is easy to control. However, the vast number of community sewage works and the waste fertilisers and animal debris from farms may be more difficult to control, especially if groundwaters, rather than surface streams, are polluted.

Already the drought years of 1975, 1976 and 1977 have resulted in high levels of nitrate affecting groundwater supplies, presumably from a source in the fertilisers used on the soil above. The exploitation of groundwater from aquifers near the coast has resulted in *saline intrusion* – salt or brackish water entering the underground strata to replace the fresh water extracted. This is a natural form of pollution but triggered by man's exploitation of the resources.

It is certain that a large proportion of our needs for drinking water could be provided by dirty water if it could be cheaply and completely cleaned up. For instance it can be used to recharge underground aquifers, becoming cleansed by percolation through the soil and rock. Alternatively we could use uncleaned water directly for our many uses which do not involve drinking – for cooling, domestic toilet flushing, agriculture and so on.

Almost certainly the next era of hydrology will see the chemist and biologist being as important as the engineer in water resources. One man's effluent is another's boating lake, fishing stream or drinking water!

Suggested further work

Experimental

Although the amounts of equipment used for 'official' hydrology are well beyond the resources of a school or college, both a basic grasp of the subject and some illuminating experimental results can be obtained very simply. One of the basic themes of what you have read has been quantities. Start by calculating what amounts of water you need yourself each day; how would you collect enough rain to supply yourself? The measurement of the rainfall in a raingauge involves some appreciation of the equation, depth x area = volume.

Having decided what volume of water you need using the table below, or by keeping a record, work out from the average rainfall for your region the area of plastic sheet you would need to collect enough rainfall.

Data for water resource calculations

average rainfall over England $2 \cdot 3$ mm per day
average rainfall over Wales $3 \cdot 7$ mm per day
average rainfall over Scotland $3 \cdot 8$ mm per day

A man requires $2 \cdot 5$ litres a day for direct consumption, plus the following amounts for life in a developed economy:
50 litres per day for flushing the lavatory
45 litres a day for washing and bathing yourself
14 litres a day for laundry
14 litres a day for washing up
 8 litres a day for the car and garden
 4 litres a day for all those extra cups of coffee and tea!

By making simple raingauges it is possible to get a better idea of local rainfall. How does rainfall vary over the area served by the school; are there urban influences or altitudinal influences?

Turning to runoff from catchments, there are many opportunities to measure small flows volumetrically. The nearest catchment we all have is a roof (if it is sloping). Building designers in fact work out the slope of roofs and the size of troughing and downpipes using hydrological analyses. Little storage of water should occur – relate the flow from your downpipe to rates of rainfall.

Other sites with small, manageable flows are tile drain systems beneath agricultural land, outfalling at the ditch. The farmer might be pleased to know how well the drains are working! One could also look at saturation levels in his soil with simple dip-wells. It might be possible for you to combine with some chemistry students to investigate the runoff from fertilisers which might be applied to the drained land.

During residential field excursions it is often feasible to install a 'V' notch weir, raingauge and possibly an evaporation pan or tin-can lysimeter on the first day, making observations at regular intervals throughout the rest of the stay. It might be possible to contact the local Water Authority for supplementary information during the period, or to work in a catchment area in which they already have equipment or have a special interest.

Remember that runoff processes are still being researched professionally using quite simple methods. New work is very welcome on interception by forests and other crops, throughflow through the soil, surface runoff and channel storage, all of which can be done using techniques described here.

Statistical

It will be unusual for schools to be able to collect long records of rainfall, river flow, soil moisture and so on. However, it may be possible to borrow data locally to get used to working with large amounts of it.

A national archive of past data is available in the form of the books called *British Rainfall*, formerly published by the Meteorological Office, *The Surface Water Yearbook*, formerly published by the Water Resources Board, and *Water Data*, published by the Department of the Environment's Water Data Unit. The latter publications cover England and Wales; the Scottish Development Department and Scottish Office hold brief reports on Scottish water statistics.

Surveys of River Pollution are also available for England, Wales and Scotland.

Further reading

The following textbooks are a sample of those available for hydrology. They have a British emphasis where possible – the abundant United States literature is not listed:

Scientific hydrology

R. C. Ward, *Principles of Hydrology*, 2nd edn., McGraw-Hill, 1975.
D. R. Weyman, *Runoff Processes and Streamflow Modelling*, Oxford University Press, 1975.

Applied hydrology

K. Smith, *Water in Britain*, 2nd edn., Macmillan Press, 1975.
M. D. Newson, *Flooding and the Flood Hazard in the United Kingdom*, Oxford University Press, 1975.
Water Resources, The Open University, 1975.

Practical work

E. W. Anderson, 'Drainage basin instrumentation in fieldwork', *Teaching Geography*, No. 21, Geographical Association, 1974.
K. J. Gregory and D. E. Walling, 'Field measurements in the drainage basin', *Geography*, **56**, 1971, pp. 277–292.
Weyman, Wilson, Cleverley and Smith, 'Hydrology for schools', *Teaching Geography*, No. 25, Geographical Association, 1975.